S0-BLB-292

John C. Eccles

How the SELF Controls Its BRAIN

Springer-Verlag
Berlin Heidelberg New York
London Paris Tokyo
Hong Kong Barcelona
Budapest

Sir John C. Eccles

Ca'a la Gra
CH-6646 Contra
Switzerland

Frontispiece: The Geographer by Jan Vermeer van Delft 1632-1675
Canvas 53.0 × 46.6 cm
owned by Städelsches Kunstinstitut, Frankfurt

With 52 Figures

ISBN 3-540-56290-7
Springer-Verlag Berlin Heidelberg New York
ISBN 0-387-56290-7
Springer-Verlag NewYork Berlin Heidelberg

CIP data applied for

Printed in Germany

Cover Design: Erich Kirchner, Heidelberg, Germany
Designed, copy-edited, and typeset by M. Seymour, St. Andrews, UK
Printing: Druckhaus Beltz, Hemsbach, Germany
Binding: Buchbinderei Schäffer, Grünstadt, Germany

SPIN 10091663 57/3140-5 4 3 2 1 – Printed on acid-free paper

Dedicated in admiration and gratitude to my dear friend Dr. Dr. h. c. mult. Heinz Götze on the wonderful occasion of his 80th birthday

Shakespearian Dualism

My brain, I'll prove the female to my soul,
My soul, the father: and these two beget
A generation of still-breeding thoughts;
And these same thoughts, people this world
In humours, like the people of this world,
For no thought is contented.

William Shakespeare, *Richard II*, Act V, Scene 5

Preface

The controversial title of this book has to be regarded as a challenging hypothesis that has been developed in the successive stages of scientific investigations described in Chapters 4–10.

It is much to be regretted that most scientific investigators of the brain are still inductivists, believing that science should be practised by accumulation of experimentally observed facts from which scientific truth is supposedly distilled. The scientific literature on the brain is an exemplar of the immense reporting of facts with a minimum of relationship to their meaning in the light of some fully expressed scientific hypothesis. Induction was shown to be untenable as a scientific method by Popper in *The Logic of Scientific Discovery* (1958). Instead, advances in scientific understanding come ideally from hypothetico-deductivism: firstly, development of a hypothesis in relation to a problem situation, and secondly, its testing in relation to all relevant knowledge and furthermore by its great explanatory power.

This is the procedure I have attempted to follow for several decades in my scientific study of *the mind–brain problem*, as described in the earlier chapters of this book. And now at the very end (Chapters 9 and 10) there has been created a hypothesis that I think merits the status of being a challenging 'how', but it still only ranks as a hypothesis in Popperian scientific method. It is of great significance that for the first time a hypothesis of the mind–brain problem has been developed in scientific detail and that it does not infringe the conservation laws of physics.

In the philosophy of the earlier decades of this century we were immersed in the long dark gloom of behaviourism, Ryleanism, logical positivism, Skinnerism, and so on. I agree with the appraisal of Roger Sperry as expressed in the extracts in Chapters 3.10 and 3.11. Since the 1950s Sperry and I have challenged this materialist interpretation of brain science, but we have been largely ignored. Materialists remain as dominant as ever because they are devotees of a dogmatic belief system which holds them with a religious-like orthodoxy, as expressed by Edelman's materialist metaphysics in Chapter 3.5.

But now entrancing gleams of light illuminate the pervasive gloom. There has been an unexpected interest in the human experience of consciousness. In Chapter 3 there are extracts specially related to consciousness from many publications by such authors as Changeux, Crick and Koch, Dennett, Edelman, Hodgson, Penrose, Searle, Sperry, and Stapp.

But one must not underestimate the materialist-monists. There has been developed by Feigl (1967) and others the strange belief in an identity theory—that mental events such as consciousness are in some manner 'identical' with brain events (see Chapter 1.3). So, with this enigmatic identity, mental states are just brain events! And materialist-monism is believed by most neuroscientists. This dominant materialism gives the brain complete superiority over the mind, even in the experience of consciousness.

A most important programme for this book is to challenge and negate materialism and to reinstate the spiritual self as the controller of the brain. We each of us have, though perhaps not fully realized, the primal certainty of the conscious self at the centre of our being, as so poetically described by Sherrington at the start of Chapter 1. The great problem confronting each of us is the manner in which our experienced self relates to our brain. This problem was first appreciated by the Greek philosophers Alcmaeon of Croton and Hippocrates, then Plato, and so through to Descartes. About a century ago William James developed a philosophy of the mind that is much appreciated by philosophers and neuroscientists today, even though the brain was very poorly understood, as was fully realized by James himself.

As I was a student of Sherrington, my life-long dedication to the mind–brain problem has been concentrated on the brain. It was to be a very long journey, a long haul, until now, when I dared with Beck to put up a hypothesis that is central to the problem, as is told in Chapter 9.

There are two wonderful concepts, 'the Self' and 'the Brain', that appear as titles of three books: (1) *The Self and Its Brain* (Popper and Eccles 1977), the title being created by Karl Popper; (2) *Evolution of the Brain, Creation of the Self* (Eccles 1989); (3) *How the Self Controls Its Brain*, the present book.

This last title is rather daring and it is meant to be so. The hypothesis that the self can effectively control the brain in intentions and attentions opens up vistas of the self controlling its brain that each of us has known from babyhood through childhood and unsophisticated adulthood. Yet it is denied by the dominant materialist philosophers.

So this book is an inexorable challenge that materialists have to answer. Can they assert that the self does not control the brain and support this denial by scientific evidence on the human neocortex in all of its subtleties? It would be expected that materialists would be in the forefront of investigations of the most advanced scientific happenings of the neocortex, but in fact they rely on the complexities of neural circuitries even for the origin of consciousness, as described by Changeux, Dennett, and Edelman in Chapter 3.

Most importantly, by using exquisite studies on the structure and function of the key microsites of the cerebral cortex (by Akert, Fleischhauer, Peters, Redman, and their associates), the hypothesis of mind–brain interaction is explained by Beck using quantum physics without infringing the conservation laws of physics (Chapter 9).

It is of great significance that the key hypothesis has been expressed in diagrammatic form in Figures 9.5 and 10.2. These figures represent the climax of the book, but on the way to this climax there are many unexpected episodes and discoveries, so that the story of 'How the Self Controls Its Brain' has all the features of a romance, the romance of my life.

Contra (TI), Switzerland, March 1993 *John C. Eccles*

Contents

Acknowledgements

I am much indebted to Dr Heinz Götze for arranging for the publication of this book by Springer-Verlag, with their excellent technical facilities.

I wish to express my thanks to the Max-Planck-Institut für Hirnforschung in Frankfurt am Main: Prof. Wolf Singer for his wise and critical advice, and particularly to Dr Manfred Klee for his generous help in many aspects of this publication, particularly in arranging for the figure publication.

The most significant and gracious acknowledgement is to my wife, Dr Helena Eccles, for her dedicated help in all stages of the creation of this book, particularly in the typing and retyping of the whole text and in her wise critical judgements. Furthermore, to her loving care I owe my survival and recovery from a severe affliction, which came on me on the night of 28 February 1993. Thus I was able to complete this book, which I feel is the greatest adventure of my life, with its spiritual dedication to the giving of hope, beauty, and meaning to life.

I wish to thank the Royal Society, the National Academy of Sciences, and *Scientific American* for their permission to reproduce figures from their publications.

I express my gratitude to the authors of publications that I have used for illustrations: Drs K. Akert, J. Szentágothai, H. Stephan, P. Roland, R. B. Kelly, A. Peters, C. Schmolke, P. S. Ulinski, D. Margoliash, H. Korn, S. J. Redman, R. J. Sayer, E. G. Gray, R. Porter, and D. Ingvar. My special thanks go to Drs M. E. Raichle and G. D. Fischbach for providing the colour slide for the four colour pictures of Figure 10.3.

Chapters 4, 5, 6, 7, and 9 are modified versions of previously published papers:

4 New light on the mind–brain problem: how mental events could influence neural events, in *Complex Systems: Operational Approaches in Neurobiology, Physics and Computers*, edited by H. Haken (Springer, Berlin, Heidelberg 1986). For the benefit of readers there is repetition of some figures and references.

5 Do mental events cause neural events analogously to the probability fields of quantum mechanics?, *Proc. Roy. Soc. London* **227**, 411–428 (1987).
6 A unitary hypothesis of mind–brain interaction in the cerebral cortex, *Proc. Roy. Soc. London B* **240**, 433–451 (1990).
7 The evolution of consciousness, *Proc. Nat. Acad. Sci.* **89**, 7320–7324 (1992).
9 (With F. Beck) Quantum aspects of brain activity and the role of consciousness, *Proc. Nat. Acad. Sci.* **89**, 11357–11361 (1992).

1 The Problem

> Each waking day is a stage dominated for good or ill, in comedy, farce or tragedy, by a 'dramatis persona', the 'self'. And so it will be until the curtain drops. This self is a unity. The continuity of its presence in time, sometimes hardly broken by sleep, its inalienable 'interiority' in (sensual) space, its consistency of view-point, the privacy of its experience, combine to give it status as a unique existence . . . It regards itself as one, others treat it as one. It is addressed as one, by a name to which it answers. The Law and the State schedule it as one. It and they identify it with a body which is considered by it and them to belong to it integrally. In short, unchallenged and unargued conviction assumes it to be one. The logic of grammar endorses this by a pronoun in the singular. All its diversity is merged into oneness.
>
> Sherrington 1951

1.1 Introduction

This poetic and vivid expression of Sherrington is a felicitous initial theme for my book, which will depend so much on his courageous but unsuccessful struggle to resolve the mystery of man (*Man on His Nature*, Sherrington 1951).

No criticism is implied by the word 'unsuccessful'. All of us who attempt to wrestle with this most intractable of problems cannot expect success in a venture that has eluded all philosophers and scientists from Aristotle to this present age. But there have been notable advances in the exploration of this vast and perplexing problem. Certainly, with our rapidly advancing knowledge of science, and particularly of biology and the neural sciences, we can anticipate that the problems can be 'seen with new eyes'. There is of course an entrenched materialist orthodoxy, both philosophic and scientific, that rises to defend its dogmas with a self-righteousness scarcely equalled in the ancient days of religious dogmatism. I, for one, derive much encouragement from this die-hard resistance. It is good to feel oneself battling against a discredited establishment! But of course I derive much more encouragement from the eminent scientists and philosophers who, each in his own way, has dared to adventure in these excessively difficult and dangerous fields of thought. For example, I refer to publications by Eddington (1939), Schrödinger (1958), Hinshelwood (1962), Kneale (1962), Beloff (1962), Wigner (1964), Dobzhansky (1967), Polanyi (1975), Thorpe (1974), Thorpe (1979), Popper (1959), Popper (1968), Popper (1972), Popper (1975), Sperry (1977), Sperry (1983), MacKay (1978), Ingvar (1975), Ingvar (1990), Granit (1977), Searle (1984), Searle (1992), Armstrong (1981), Creutzfeldt and Rager (1978), Szentágothai (1975), and Szentágothai (1979).

Following Popper (1968) I can say:

> I wish to confess, however, at the very beginning, that I am a realist: I suggest somewhat like a naive realist that there is a physical world and a world of states of consciousness, and that these two interact.

I have long been concerned with a problem that has been expressed very succinctly by Schrödinger (1958):

> The world is a construct of our sensations, perceptions, memories. It is convenient to regard it as existing objectively on its own. But it certainly does not become manifest by its mere existence. Its becoming manifest is conditional on very special goings-on in very special parts of this very world, namely on certain events that happen in a brain. That is an inordinately peculiar kind of implication, which prompts the question: what particular properties distinguish these brain processes and enable them to produce the manifestation?

1.2 Hypotheses Relating to the Mind–Brain Problem

It is not possible here to give a detailed appraisal of the immense philosophical literature on the mind–brain problem, or the body–mind problem. Fortunately this has been done in a masterly manner by Popper (Popper and Eccles 1977, Chapters P1 and P3–P5). He has critically surveyed the historical development of the problem from the earliest records of Greek thought. I will begin with a simple description and diagram of the principal varieties of this extremely complex and subtle philosophy, concentrating specifically on the formulations that relate to the brain rather than the body, because clinical neurology and the neurosciences make it abundantly clear that the mind has no direct access to the body. All interactions with the body are mediated by the brain, and furthermore only by the higher levels of cerebral activity.

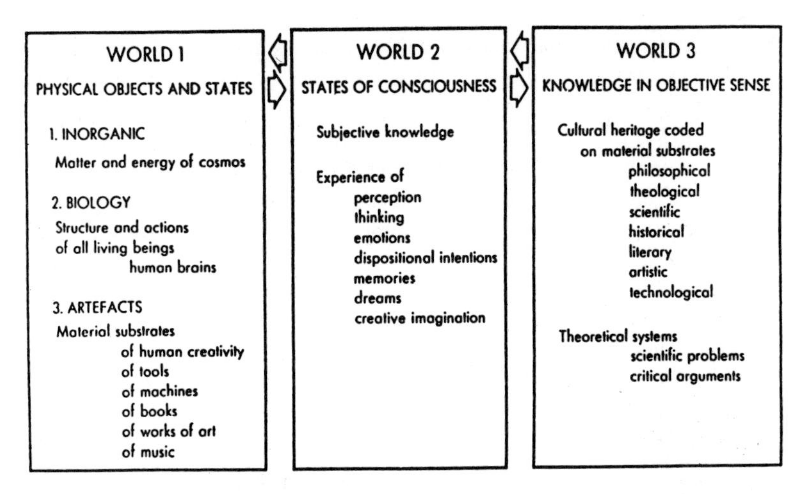

Figure 1.1 A tabular representation of the three worlds that comprise all existents and all experiences, as defined by Popper.

Diagrammatic representation of brain–mind theories

World 1 = All of material or physical world including brains
World 2 = All subjective or mental experiences
World 1_P is all the material world that is without mental states
World 1_M is that minute fraction of the material world with associated mental states

Radical Materialism:	World 1 = World 1_P; World $1_M = 0$; World 2 = 0.
Panpsychism:	All is World 1–2, World 1 or 2 do not exist alone.
Epiphenomenalism:	World 1 = World 1_P + World 1_M World $1_M \rightarrow$ World 2
Identity theory:	World 1 = World 1_P + World 1_M World 1_M = World 2 (the identity)
Dualist – Interactionism:	World 1 = World 1_P + World 1_M World $1_M \rightleftarrows$ World 2; this interaction occurs in the liaison brain, LB = World 1_M. Thus World 1 = World 1_P + World 1_{LB}, and World $1_{LB} \rightleftarrows$ World 2

Figure 1.2 A schematic representation of the various theories of brain and mind.

For our present purpose it is of value to clarify the arguments (Figure 1.1) by developing an explanatory diagram (Figure 1.2) of the principal theories so that the materialist theories of the mind can be contrasted with the dualist-interactionist theory that is being proposed here (Figure 1.2).

- *World 1* is the whole material world of the cosmos, both inorganic and organic, including all of biology, even human brains, and all man-made objects.
- *World 2* is the world of conscious experiences, or of the mind, not only of our immediate perceptual experiences, visual, auditory, tactile, pain, hunger, anger, joy, fear, etc., but also of our memories, imaginings, thoughts, planned actions, and centrally thereto of our unique self as an experiencing being.
- *World 3* is the world of human creativity—for example, the objective contents of thoughts underlying scientific, artistic, and literary expression. Thus World 3 is the world of culture in all of its manifestations, as has been expressed by Popper (Popper and Eccles 1977, Chapter P2).

1.3 Materialist Theories of the Mind–Brain Problem

The dominant theories of the mind–brain relationship that are today held by neuroscientists are purely materialistic in the sense that the brain is given complete mastery (Pribram 1971; Rensch 1971, 1974; Barlow 1972; Doty 1975; Blakemore 1977; Mountcastle 1978; Edelman 1978, 1989; Changeux 1985).

The existence of mind or consciousness is usually not denied, but it is relegated to the passive role of mental experiences accompanying some types of brain action, as in psychoneural identity, but with absolutely no *effective* action on the brain. The complex neural machinery of the brain functions in its determined materialistic fashion regardless of any consciousness that may accompany it. The 'common sense' experiences that we can control our actions to some extent or that we can express our thoughts in language are alleged to be illusory. An effective causality is denied to the self-conscious mind per se.

In Figure 1.2 World 1 is divided into World 1_P and an infinitessimally small World 1_M. In general, materialist theories are those subscribing to the statement that mental events can have no *effective* action on the brain events in World 1—that World 1 is closed to any conceivable outside influence such as is postulated in dualist-interactionism. This *closedness* of World 1 is ensured in four different ways in the four varieties of materialism illustrated in Figure 1.2.

1.3.1 Radical Materialism

There is a denial of the existence of conscious processes and mental states. Radical behaviourism provides a complete explanation of behaviour, including verbal behaviour and the dispositional states that lead to this. I do not think that many neuroscientists hold such an extreme view. It is, however, attractive to some philosophers because of its simplicity and because it eliminates not only the mind–brain problem, but also the problem of the origin of mind. The cosmos is reduced to the pristine simplicity that it had before the origin of life and of mind. While this may appeal to reductionist philosophers, neuroscientists must find this extreme reductionism absurd, so it will not be further discussed. A comprehensive discussion leading to rejection can be found in Popper and Eccles (1977, Section 18).

1.3.2 Panpsychism

This is a very ancient theory developed by the earliest Greek philosophers, who proposed that 'soul is mingled with everything in the whole universe'. Such philosophers as Spinoza and Leibniz espoused various forms of panpsychism. Essentially their belief was that all things had an inner psychical aspect and were material in their outer aspect. Panpsychism has even attracted modern biologists such as Waddington (1969), Rensch (1971), and Birch (1974), because it offers such an attractive solution to the problem of the evolutionary origin of consciousness, namely that consciousness was associated with all matter in some protopsychic state and was merely developed with the increasing complexity of the brain to appear as the self-consciousness associated with the human brain. There is a parallelistic association of consciousness with brain events in the manner of the inner and outer aspects of an egg shell! However, modern physics does not admit memory or identity to elementary particles—electrons, protons, neutrons—and thus the panpsychist doctrine of the 'protoconsciousness' of such particles must be rejected. Panpsychism thus does not escape the problem of the origin of consciousness. Both panpsychism and radical materialism have the attraction that the universe is monistic and homogeneous. But we shall see that the cost of this simplicity is prohibitive (see also Popper and Eccles 1977, Part I, Section 19).

1.3.3 Epiphenomenalism

Epiphenomenalism differs from panpsychism in that mental states are attributed only to animals that exhibit mind-like behaviour, such as learning and reacting intelligently and purposively. All varieties of epiphenomenalism have as a central tenet the thesis that the mental processes are completely ineffective in controlling behaviour. The neural machinery works without any influence from consciousness, just as, according to T. H. Huxley, the work of a steam locomotive is uninfluenced by the sound of the steam whistle! Yet it is proposed that at a certain stage of evolution these ineffective mental states emerged and were then greatly developed in the evolutionary process to full human self-consciousness. The actual relationship of the mind–brain influence is not defined, but one view is that mind states run parallel with brain states, much as in panpsychism (see Popper and Eccles 1977, Part I, Section 20).

1.3.4 The Psychophysical Identity Theory, or the Central State Theory

Like panpsychism this theory was first developed by Greek philosophers, and the two theories have often been linked, for example by Spinoza and Rensch. The most subtle and acceptable form of the theory has been given by Feigl (1967). There are many variants. Several analogies have been used in illustrating the postulated identity, but all are unsatisfactory because both components are in the materialist mode. For example, there is the much overworked analogy of the evening star and morning star achieving identity in the planet Venus. Other analogies are: cloud and fog, achieving identity in water droplets of the atmosphere; or a flash of lightning = electric discharge; or genes = DNA.

Nevertheless there are attractive and important features in the identity theory. Mental processes are regarded as real or things in themselves. They are conjectured to be a property of a very small and select group of material objects, namely neural events in the brain, and probably in special regions of the brain. The conscious experiences are known from within—*knowledge by acquaintance*—whereas the 'identical' physical events are known from without by description—*knowledge by description*—of the neural events in the brain. These events described by the neuroscientist turn out to be the experiences consciously perceived. Thus the key postulate is essentially a parallelism or an inner and outer aspect. It is surprising that there was initially so little in the way of developing this challenging hypothesis, particularly in attempts to identify the neural events distinguished by this identity criterion, not only anatomically but also physiologically; but now this has been attempted (Changeux 1985; Edelman 1989)(see Chapters 3.1 and 3.5). Polten (1973) has critically analysed the identity theory as defined by Feigl (1967). This attack on logical grounds is severe and comprehensive and is as yet apparently unanswered. For further discussion see Popper and Eccles (1977, Part I, Section 22).

Neuroscientists find the identity theory attractive because it gives the future to them (Changeux 1985; Edelman 1989)(see Chapters 3.1 and 3.5). It is admitted that our present understanding of the brain is quite inadequate to provide more than a crude explanation of how the brain provides all the richness and wonderful variety of perceptual experiences, or how the mental events or thoughts can have the immense range and fruitfulness that our imaginative insights achieve in their action on the world.

However, all this is taken care of by the theory that has been named promissory materialism (Popper and Eccles 1977, Part I, Section 26). This theory derives from the great successes of the neurosciences, which undoubtedly are disclosing more and more of what is happening in the brain in perception, in the control of movement, and in states of consciousness and unconsciousness. The aim of these research programmes is to give a more and more complete and coherent account of the manner in which the total performance and experience of an animal and of a human being are explicable by the action of the neural machinery of the brain. According to promissory materialism this scientific advance will progressively restrict the phenomena that appear to require mental terms for their explanation, so that in the fullness of time everything will be describable in the materialist terms of the neurosciences. The victory of materialism over mentalism will be complete. I regard this theory as being without foundation. The more we discover scientifically about the brain the more clearly do we distinguish between the brain events and the mental phenomena and the more wonderful do the mental phenomena become. Promissory materialism is simply a superstition held by dogmatic materialists. It has all the features of a Messianic prophecy, with the promise of a future freed of all problems—a kind of Nirvana for our unfortunate successors. In contrast the true scientific attitude is that scientific problems are unending in providing challenges to attain an even wider and deeper understanding of nature and man.

An outstanding characteristic of identity theories that centre around Feigl's (1967) brilliant formulation is the multiplication of names for theories. There are, for example: emergent interactionism (Sperry 1977, 1980); identistic panpsychism (Rensch 1974); physicalism (Smart 1963); biperspectivism (Laszlo 1972); emergentistic materialism (Bunge 1980).

1.3.5 General Discussion of Materialist Theories

I propose now to consider the biological implications of the three materialist theories that admit the existence of consciousness or mental states (World 2). Despite the differences in detail with respect to the relationship of World 2 to World 1, all are in agreement that the physical events in the brain (World 1) are alone causally effective in bringing about actions. In panpsychism the mental accompaniments of brain events are given no more causal effectiveness than in epiphenomenalism. They are merely the necessary concomitants of the on-going brain activities. At first it seems otherwise with the identity theory, where World 1_M can interact with World 1_P because both are components of the neural machinery in the brain. Thus we have: World $1_P \rightleftharpoons$ World 1_M and World 1_M relates to World 2. Nevertheless the performance of the brain in controlling behaviour is entirely within the physical structures of the brain. No causal effectiveness of World 2 is admitted other than that of pertaining to World 1_M. Thus the closedness of World 1 is as absolute as with panpsychism or epiphenomenalism.

These three theories assert the causal ineffectiveness of World 2. Yet this is an undeniable fact (see Popper and Eccles 1977, Part I, Sections 20 and 23, Dialogue VIII; and Chapters 5–10 of the present book). There is firstly its emergence and then its progressive development with the growing complexity of the brain. In accord with evolutionary theory only those structures and processes that significantly aid in survival are developed in natural selection. If World 2 is impotent, its development cannot be accounted for by evolutionary theory. It has not been recognized by the proponents of panpsychism, epiphenomenalism, and the identity theory that they are advocating a theory that is unrelated to the theory of biological evolution. According to that theory mental states and consciousness (World 2) could have evolved and developed only if they were *causally effective* in bringing about changes in neural happenings in the brain with the consequent changes in behaviour having survival value. That can occur only if World 1 of the brain is open to influences from the mental events of World 2, which is the basic postulate of the dualist-interactionist theory.

1.4 The Dualist-interactionist Theory

This theory is the most ancient formulation of the mind–brain problem, being in some form generally accepted by Greek thinkers from Homer onwards (Popper and Eccles 1977, Part I, Sections 43 and 46). It was developed by Descartes, who attempted to define a detailed mode of operation that led to it being rejected in favour of some form of parallelism. In its modern form it is distinguished from all parallelistic theories precisely by the requirement of the openness of World 1 to World 2 events (Figure 1.3).

The essential feature of dualist-interactionism is that the mind and brain are independent entities, the brain being in World 1 and the mind in World 2, and that they interact by quantum physics, as will be described in Chapters 9 and 10 and as illustrated in Figures 9.5 and 10.2.

There is a frontier, and across this frontier there is interaction in both directions, which can be conceived as a flow of information, not of energy. Thus we have the extraordinary doctrine that the world of matter–energy (World 1) is not completely sealed, which is a fundamental tenet of classical physics, but that there are subtle communications in what is otherwise the completely closed World 1. On the contrary, the closedness of World 1 has been safeguarded with great ingenuity in all materialist theories of the mind.

None of the materialist theories affirms the dualistic concept of the world of mind (World 2) interacting with the brain (World 1) that is open to it.

1.5 Critical Evaluation of Mind–Brain Hypotheses

Great display is made by all varieties of materialists that their brain–mind theory is in accord with natural law as it now is. However, this claim is invalidated by two most weighty considerations.

Firstly, nowhere in the laws of physics or in the laws of the derivative sciences, chemistry and biology, is there any reference to consciousness or mind. Shapere (1974) makes this point in his strong criticisms of the panpsychist hypothesis of Rensch (1974) and Birch (1974), in which it was proposed that consciousness or protoconsciousness is a fundamental property of matter. Regardless of the complexity of electrical, chemical, or biological machinery there could be no statement that there is an emergence of this strange non-material entity, consciousness, or mind. This is not to affirm that consciousness does not emerge in the evolutionary process but merely to state that its emergence is not reconcilable with the laws of classical physics, which are the natural laws as at present understood (see Chapters 3.12–14 on Stapp). For example, such laws do not allow any

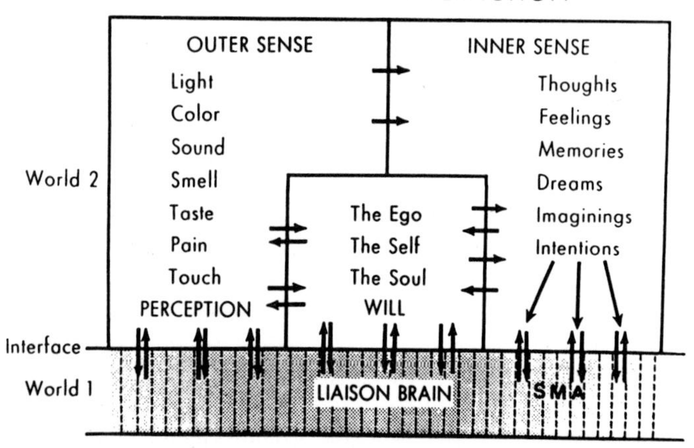

Figure 1.3 An information flow diagram for brain–mind interaction. The three components of World 2, outer sense, inner sense, and the ego, or self, are diagrammed with their communications shown by *arrows*. Also shown are the lines of communication across the interface between World 1 and World 2, that is, from the liaison brain to and from these World 2 components. The liaison brain has the columnar arrangement indicated by the *vertical broken lines*. It must be imagined that the area of the liaison brain is enormous, with open modules numbering over a million, not just the two score depicted here.

statement that consciousness emerges at a specified level of complexity of systems, which is gratuitously assumed by all materialists (Changeux 1985; Edelman 1989; Searle 1984, 1992)(see Chapters 3.1, 3.5, 3.8, and 3.9) except panpsychists. Their belief that some primordial feelings attach to all matter, presumably even to atoms and subatomic particles (Rensch 1971), finds no support whatsoever in physics. One can also recall the poignant questions by computer-lovers. At what stage of complexity and performance can we agree to endow computers with consciousness? Mercifully this emotionally charged question need not be answered. You can do what you like to computers without qualms of being cruel!

Secondly, as stated above, all materialist theories of the mind are in conflict with biological evolution. Since they all (panpsychism, epiphenomenalism, and the identity theory) assert the causal ineffectiveness of consciousness per se, they fail completely to account for the biological evolution of consciousness (Chapter 7), which is an undeniable fact. According to biological evolution, mental states and consciousness could have evolved and developed *only if they were causally effective* in bringing about changes in neural happenings in the brain with the consequent changes in behaviour.

This can occur *only if* the neural machinery of the brain is open to influences from the mental events of the world of conscious experiences, which is the basic postulate of dualist-interactionist theory.

Finally, the most telling criticism of all materialist theories of the mind is against its key postulate that the happenings in the neural machinery of the brain provide *a necessary and sufficient explanation of the totality both of the performance and of the conscious experience of a human being*. For example, the willing of a voluntary movement is regarded as being *completely determined* by events in the neural machinery of the brain, as also are all other cognitive experiences. But as Popper states (Popper 1972, Chapter 6):

> . . . physical determination is a theory which, if it is true, is unarguable since it must explain all our reactions, including what appear to us as beliefs based on arguments, as due to purely physical conditions. Purely physical conditions including our physical environment make us say or accept whatever we say or accept.

This is an effective *reductio ad absurdum*. This stricture applies to all of the materialist theories of the mind. But now in Chapter 9 of the present book the application of quantum physics to the ultramicrostructure and function of the brain (neocortex) reveals that mental action could increase the neural responses by increasing the quantum probability of exocytosis, without interfering with the conservation laws of physics (Beck and Eccles 1992 and Chapter 9). So dualist-interaction can now be explained in principle.

2 Dualist-interactionism—My Story

This chapter relates to a philosophy that has been integral with my intellectual life since my early years as a medical student. I had been confronted by an irreligious philosophy of monist-materialism that I could not accept. In my wide reading of philosophy from the pre-Socratics to the present, I had come to the dualist philosophy of Descartes with his *res externa* and *res cogitans*. It gave a secure status to the human soul or self, but his interaction story was completely erroneous. In fact from my reading of philosophy I discovered a pervasive ignorance of the brain at the subtle level at which it could relate to conscious experiences, for example, in neurology. So I decided to concentrate my studies on the brain. My initial success was to go in 1925 from Melbourne to Oxford as a Rhodes Scholar to work with the greatest neuroscientist of the age, Sir Charles Sherrington.

Soon I recognized that my task was to be a life-long study of the brain, or neuroscience as it came to be called!

My dualist-interactionist philosophy was like a beacon light guiding my way through the complexities of my neuroscientific studies. There were episodes of philosophy where I was able to utilize my intimate scientific knowledge of the brain. In 1975 I retired from experimental work and devoted myself more and more to the philosophy of the brain. In 1977 there came my greatest opportunity in my collaboration with Sir Karl Popper: *The Self and Its Brain—An Argument for Interactionism*, which has become recognized as giving the strongest presentation of dualist-interactionism.

My first philosophical publication was in *Nature* in 1951: 'Hypotheses Relating to the Brain–Mind Problem'. It arose from a lecture I gave at the invitation of Professor Passmore to the philosophy department of the University of Otago, Dunedin, which was much criticized by the assembled materialist philosophers who knew almost nothing on the brain. The whole episode was strangely recounted by Passmore after a lecture I gave to the Royal Belgian Academy more than 20 years later!

2.1 'Hypotheses Relating to the Brain–Mind Problem', *Nature*, 1951

In my effort to imagine how the cerebral cortex could relate to mind, I had perforce to concentrate on the organization of the cortical neurons in complex functional networks with synapses as generators of neuronal discharges. The patterns of neuronal activity would be due to the very recent synaptic inputs on to the patterns of structural specificity, both inherited and acquired.

Cartesian dualism necessarily involved the problem of how mind and brain could interact in willed acts and in perception. It had been recognized that consciousness involved the liaison of mind with brain, and the electroencephalogram showed that the cerebral cortex had to be very active in a selective way. The intense driven activity of an epileptic seizure led to an immediate loss of consciousness. These concepts relate the spatio-temporal patterns in the neuronal networks to the specific experiences in the mind but say nothing concerning the 'how' of that linkage. That has been my ongoing problem. With memory it is proposed that the remembered thought appears in the mind as the spatio-temporal pattern is being replayed in the cortex.

During inattention with eyes closed the dominant alpha rhythm in the EEG indicates an absence of specific action patterns. With active perception and attention there develop activity patterns in the EEG, indicating the formation of specific spatio-temporal patterns in the neuronal net. Sherrington (1940) vividly and aptly pictures the neuronal net as an

> enchanted loom where millions of flashing shuttles (the nerve impulses) weave a dissolving pattern, always a meaningful pattern, though never an abiding one, a shifting harmony of sub-patterns.

Unfortunately the neuronal structure of the cerebral cortex was poorly understood in 1951, so that it was only possible to consider the neuronal networks in general terms. The hypothesis was proposed that the mind achieves liaison with the brain by exciting spatio-temporal 'fields of influence' that become effective through the unique detector function of the specific neuronal nets that are tuned in to the mind influences.

2.2 *The Neurophysiological Basis of Mind,* 1953

The invitation to give in 1952 the Waynflete Lectures at Oxford University ('The Mind–Brain Problem', Eccles 1953, Chapter 8) gave me the opportunity to develop and modify my dualist hypothesis on that problem after a full treatment of the cerebral cortex (Eccles 1953, Chapter 7).

In Chapter 7 of that book it appeared that the cortical neurons in themselves with their synapses were not fundamentally different from neurons elsewhere in the brain. The differences in performance apparently arose from complexities of their connections in neuronal nets, which Sherrington likened to an enchanted loom as already quoted. Sherrington (1951) summed up the problem well:

> Every nerve-cell of the millions in [the cerebral cortex] is clearly at a glance a nerve-cell. But nerve-cells as a class are elsewhere not specially concerned with mind. It is partly conjecture whether the properties of all these nerve-cells, their fibres, their cell-contacts (synapses), their cell-bodies, have rigidly those characters observed in the more accessible cells of the spinal cord and elsewhere. That the properties will not differ fundamentally from those elsewhere seems safe to suppose.

In the same Chapter 7 a detailed description of all the investigations on the cerebral cortex did not disclose a challenge to Sherrington's supposition.

Sherrington (1951) refers to mind as the non-sensual concept, saying:

> Our mental experience is not open to observation through any sense organ . . . It has no such channel of entrance to the mind . . . The mental act of 'knowing' . . . is experienced not observed.

This experience of self-consciousness achieves public status by symbolic communication between experiencers, especially through language. It may therefore be revealed as a fact of experience. Similarly we can report our mental experiences to others and discover that they have like experiences to report to us. In that way any hallucinations can be recognized and rejected.

Another word that needs comment is 'self'. It will be used to connote an experienced unity that derives from a linking by memory of conscious states that are experienced at widely different times—spread over a lifetime. Thus, in order that a 'self' may exist there must be some continuity of mental experiences and, particularly, continuity bridging gaps of unconsciousness. For example, the continuity of our 'self' is resumed after sleep, anaesthesia, and the temporary amnesias of concussion and convulsions.

The Cartesian dualism of mind and matter necessarily involved the problem of how mind and brain could interact in perception and in willed acts. It is easy of course to discredit Descartes's explanation of this interaction because at that time scientific knowledge of the brain was virtually non-existent and he developed a crude mechanical explanation. It has been my task to discover the scientific answer to this problem, and this has just been described in a conjoint publication (Beck and Eccles 1992 and Chapter 9).

A key question is: How does it come about that liaison with mind occurs only in special states of the matter–energy system of the cerebral cortex? Is there any special property of this system that places it in a separate category from all the remainder of the matter–energy, or natural world, and even from its own matter–energy system when it is not in this special state of activity? It is here hypothesized that such a special property in outstanding measure is exhibited by the dynamic patterns of activity in the neuronal networks that occur in the cerebral cortex during conscious states, and the hypothesis is developed that the brain by means of this special property enters into liaison with mind, having the function of a 'detector' that has a sensitivity of a different kind and order from that of any physical instrument.

The neurophysiological hypothesis of will is that 'will' modifies the spatio-temporal activity of the neuronal network by exerting spatio-temporal 'fields of influence' that become effective through this unique detector functon of the active cerebral cortex. It will be noted that this hypothesis assumes that the 'will' or 'mind influence' has itself a spatio-temporal patterned character in order to allow it this operative effectiveness.

In perception the usual sequence of events is that some stimulus to a receptor organ causes the discharge of impulses along afferent nerve fibres, which, after various synaptic relays, eventually evoke specific spatio-temporal patterns of impulses in the neuronal network of the cerebral cortex. The transmission from receptor organ to cerebral cortex is by a coded pattern that is quite unlike the original stimulus, and the spatio-temporal pattern evoked in the cerebral cortex would be again different. Yet, as a consequence of this cerebral pattern of activity, we experience sensations (more properly the complex constructs called percepts) which are 'projected' to somewhere outside the cortex; it may be to the surface of the body or even within it, or, as with visual, acoustic, and olfactory receptors, to the outside world. However, the only necessary condition for an observer to see colours, hear sounds, or experience the existence of his own body is that appropriate patterns of neuronal activity shall occur in appropriate regions of his brain, as was first clearly seen by Descartes. It is immaterial whether these events are caused by local stimulation of the cerebral cortex or some part of the afferent nervous pathway, or whether they are, as is usual, generated by afferent impulses discharged by receptor organs. Personal experiment from earliest childhood onwards and communication with other observers are the standard procedures by which we learn to interpret a part of our private perceptual experiences as events in a single physical world common to other observers.

But the key problem in perception has so far remained beyond this discussion. We may ask: How can some specific spatio-temporal pattern of neuronal activity in the cerebral cortex evoke a percept in this mind?

We shall return to this perplexing and mystifying problem towards the end of the book.

In conclusion it should be pointed out in developing the philosophy of dualist-interaction that attention has been directed firstly to the neuronal networks of the cerebral cortex that in neuroscience are considered to operate according to classical physics with nerve impulses, synaptic transmission, etc. In conscious states the cerebral cortex has been postulated to be in a state of extreme sensitivity, as a detector of minute spatio-temporal fields of influence. The hypothesis is developed that these spatio-temporal fields of influence are exerted by the mind on the brain in willed action. However, this hypothesis is still extremely problematical. For example, there was no concept of the nature of the mind that could exert these 'ghost-like influences' and of its presumed spatio-temporal patterning that gives its specific relationship to the events in the neuronal networks. Chapter 9 describes how the situation was transformed in 1992.

The great attraction of dualist-interactionism is that it gives a central role to the self, which is our fundamental experience. Nevertheless, I gave up philosophy for about 20 years and did not attempt a serious involvement until there had been great advances in the neuroscience of the cerebral cortex.

2.3 My Long Interregnum, 1952–1969

Following my 1953 book there was a long philosophic interval. I still believed in dualist-interaction and in the uniqueness of the self. However, my interests were concentrated on neuroscience, which provided the scientific challenge of my life, particularly the 14 years at the Australian National University at Canberra (1952–1966). I did not engage in philosophical adventures on dualist-interaction, but I accepted invitations to lecture and write on the mind–brain problem, which had remained much as I left it in 1952. Also experimentally I left the spinal cord and brain stem for the cerebral cortex, the hippocampus. Then in 1963 I made a great departure to the cerebellum, from 1963 to 1976, wryly remarking that in the cerebellum I could study neuroscience without being involved in the possibility of disturbance by conscious phenomena!

2.4 Brain and Conscious Experience

In 1964 there was an attractive opportunity to return to the brain and consciousness when I organized an international symposium in the Pontifical Academy of Science in Rome, under the title 'Brain and Conscious Experience'. There was a wonderful gathering of brain scientists, but our philosophical performance was disappointing. I very much missed Karl Popper.

In Chapter 2 of the Symposium, published by Springer-Verlag (Eccles 1966), I brought together all the scientific findings on the neurons of the cerebral cortex, which had been studied using the most advanced techniques by Phillips, Creutzfeldt, Lux, Klee, Armstrong, Li, and Purpura. It was remarkable to corroborate the suggestion of Sherrington quoted above that cortical neurons share in the same properties as other neurons. So the construction of neural networks has good experimental support even to the details of neuronal performance.

However, it was not possible to advance further in the attempt to understand how mind and brain could interact. The belief of strong dualism that brain received from conscious mind in willed actions and transmitted to mind to give conscious experiences had to await new scientific advances. More encouraging were the several good studies on the uniqueness of the self, for example by Bremer, Libet, Penfield, Adrian, Jasper, Moruzzi, Thorpe, and MacKay.

2.5 My Philosophical Renaissance, 1969 Onwards

There was a gradual recovery over many years with no outstanding features until 1984.

I chose 1969 as my starting point here because I was invited to give the Foerster Lecture at Berkeley with the remarkable title 'The Brain and the Soul'.

My lecture drew the largest audience I have ever had. The assigned lecture room for 2 000 was completely filled, so we had the initial disorder of moving to the largest audience room in Berkeley with over 3 000 places.

I was a bit apprehensive of the title, wondering if it was chosen to lead me on! I accepted it as a challenge. In the past I had been reluctant to use the religious word, soul, greatly preferring the philosophical word, self. In my lecture also I was careful to establish my academic credentials by building up the anatomical and histological structure of the brain and also the neuronal machinery of the cerebral cortex with the neuronal nets, states of consciousness, and freedom of the will. Then came the 3-world concept of Popper, with special emphasis on cultural evolution with language, and then the concept of the self, with self-awareness and death-awareness. Only then did I come to the soul:

> I believe that my experiencing self is only in part explained by the evolutionary origin of my body and brain, that is, of my World-1 component. It is a necessary, but not a sufficient, condition. About the origin of our world of conscious experience (World 2) [Figure 1.1] it can be described as having an emergent relation to the evolutionary development of the brain. The uniqueness that I experience cannot be attributed to the uniqueness of my genetic inheritance, as I have already argued. Our coming-to-be is as mysterious as our ceasing-to-be at death. Can we therefore not derive hope because our ignorance about our origin matches our ignorance about our destiny? Cannot life be lived as a challenging and wonderful adventure that has meaning to be discovered?

There was an enthusiastic discussion for over an hour, so I felt satisfied with my reappearance. There were two remarkable aftermaths. On the next morning (1 May) there began on Berkeley Campus the student riots that swept from there to much of America, and lasted for years. How was my lecture related to that? The second aftermath was that I sent the manuscript for publishing by Berkeley University Press as was required for the Foerster Lecture. After some weeks it was rejected because the Press was too busy! So the materialists got me! Fortunately I was assembling some papers to publish by my good friend Heinz Götze. This Foerster Lecture forms the most significant component as Chapter 10 of *Facing Reality* (1970).

These collected lectures show that even before 1970 I had returned to the philosophy of the mind–brain problem and the human self.

My next significant advance was at the conference 'Studies in the Philosophy of Biology' (Ayala and Dobzhansky 1972). There were 16 distinguished participants. My paper was entitled 'Cerebral Activity and Consciousness'. There had been great advances in technique, but the conclusion was much the same as before. The neurophysiological hypothesis is that the 'will' modifies the spatio-temporal activity of the neuronal network by exerting spatio-temporal 'fields of influence' that become effective through this unique detector function of the active cerebral cortex. It will be noted that this hypothesis assumes that the 'will' or 'mind influence' has itself some spatio-temporal patterned character in order to allow it this operative effectiveness. The only scientific advance had been that the mind influence of willing had been shown to have the postulated action on the brain as revealed in the readiness potential of Deecke and Kornhuber (1978). But still the 'how' of that action eluded us.

The inestimable advantage of this conference was that Popper and I realized that we were together in our attempt to develop the dualist-interaction concept. Karl actually asked me to join him in writing a book, he to do the philosophy and I the neuroscience. In fact we arranged to begin the collaboration in a month-long stay at the Villa Serbelloni in September 1974, where our conference was being held.

An important new advance was with Szentágothai's superb studies of the histological structure of the cerebral cortex, as illustrated in Figures 2.1b, 6.5, 7.1, and 9.1, which show the dominance of the pyramidal neurons in the building up of neuronal patterns of a cortical module.

> As yet we have little knowledge of the inner dynamic life of a module, but we may conjecture that, with its complexly organized and intensely active properties, it could be a component of the physical world (World 1) that is open to self-conscious mind (World 2) both for receiving from and for giving to [Figure 1.1]. We can further propose that not all modules in the cerebral cortex have this transcendent property of being 'open' to World 2, thus being the World 1 components of the interface. (Popper and Eccles 1977, Chapter E7)

It must be assumed that there are dynamic patterns in the development and interaction of modules, as is simply illustrated in Figures 2.2, 6.5, and 6.11.

> Thus it is proposed that the self-conscious mind exercises a superior interpretive and controlling role upon the neural events by virtue of a two-way interaction across the interface between World 1 and World 2 [Figure 1.3]. It is proposed that the unity of conscious experience comes not from an ultimate synthesis in the neural machinery, but in the integrating action of the self-conscious mind on what it reads out from the immense diversity of neural activities in the liaison brain [Figure 1.3].

> An attempt was made to show how the operative features of modules of the cerebral cortex could result in properties of such subtlety that they could be recipients of the weak actions that are postulated to be exerted by the self-conscious mind across the interface. These actions are evidenced by voluntary movements and by the recall of memories.

In Dialogue V of Popper and Eccles (1977) there is expressed a new insight into the mind–brain problem:

> Let us develop the hypothesis that the self-conscious mind is not just engaged passively in reading out the operation of neural events, but that it is an actively searching operation. There is displayed or portrayed before it from instant to instant the whole of the complex neural processes, and, according to attention and choice and interest or drive, it can select from this ensemble of performances in the liaison brain, searching now this, now that, and blending together the results of readouts of many different areas in the liaison brain. In that way the self-conscious mind achieves a unity of experience.

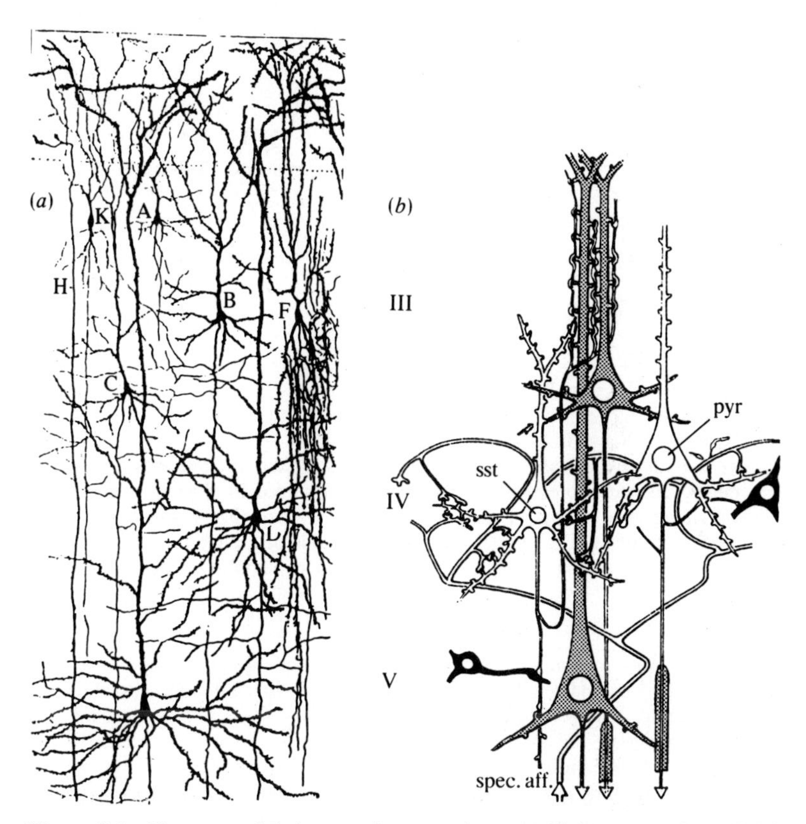

Figure 2.1 Neurons and their synaptic connections. (**a**) Eight neurons from Golgi preparations of the three superficial layers of frontal cortex from a month-old child. Small (B, C) and medium (D, E) pyramidal cells are shown with their profuse dendrites covered with spines. Also shown are three other cells (A, F, K), which are in the general category of Golgi type II with their localized axonal distributions (Ramón y Cajal 1911). (**b**) The direct excitatory neuron circuit of the specific (sensory) afferent (spec. aff.). Both spiny stellate (sst) with ascending main axon and apical dendrites of both lamina III and IV pyramidal cells (*stippled*) are probably the main targets. (Szentágothai 1979)

The mind–brain interaction was treated at several places in my Gifford Lectures, 'The Human Mystery' (Eccles 1979) and 'The Human Psyche' (Eccles 1980), but the main theme was concentrated on the great philosophical issues of our nature and our destiny with the scope of natural theology.

However, I did attempt to go further in developing the hypotheses of the self-conscious mind in relation to the brain:

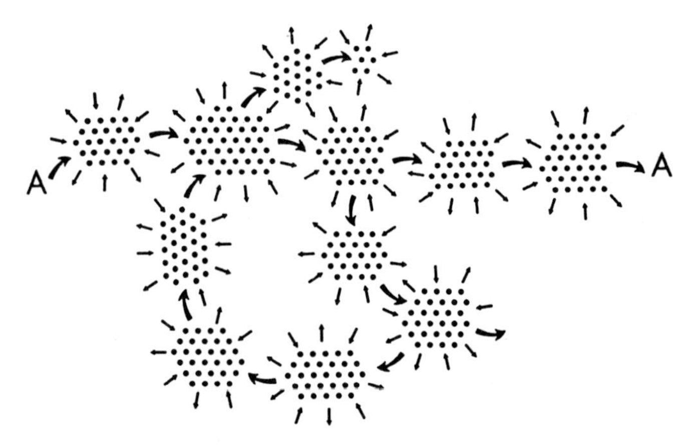

Figure 2.2 In this schema of the cerebral cortex looked at from above, the large pyramidal cells are represented as *dots* that are arranged in clusters, each cluster corresponding to a column or module. The *arrows* symbolize impulse discharges along hundreds of lines in parallel, which are the mode of excitatory communication from column to column. Only a minimal system of serially excited columns is shown.

> The self-conscious mind is actively engaged in reading out from the multitude of liaison modules [Figure 1.3] that are largely in the dominant cerebral hemisphere. The self-conscious mind selects from these modules according to attention and interest, and from moment to moment integrates its selection to give unity even to the most transient experiences. Furthermore the self-conscious mind acts upon these modules, modifying their dynamic spatio-temporal patterns. Thus it is proposed that the self-conscious mind exercises a superior interpretative and controlling role. A key component of the hypothesis is that the unity of conscious experience is provided by the self-conscious mind, not by the neural machinery of the liaison areas of the cerebral hemisphere. Hitherto it has been impossible to develop any neurophysiological theory that explains how a diversity of brain events comes to be synthesized so that there is a unified conscious experience of a global or gestalt character. The brain events remain disparate, being essentially the individual actions of countless neurons that are built into modules and so participate in the spatio-temporal patterns of activity. The brain events provide no explanation of our commonest experience, the visual world seen as a global entity from moment to moment. (Eccles 1979)

Despite the impressive description of the complex modular operations of the cerebral cortex, it was not possible to give any answer to the question of the 'how' of this operation. Neuronal complexity was not enough, even with the transcendent operations of classical physics. So I lived on into the 1980s and beyond my 80th birthday with my life-long quest unsatisfied.

Meanwhile I was concentrating on what seemed to be the key study on mind–brain interaction, voluntary movements, particularly in the key role of the part of the cerebral cortex called the supplementary motor area. In 1984 I helped to organize a conference especially dedicated to that theme: 'How the Self Operates on the Brain', Schloss Ringberg, 1984.

There was general agreement that in voluntary movement the self could bring about neural events, but the 'how' of this happening still eluded us. The monist materialists used the conservation laws of physics to contradict such happenings.

2.6 Quantum Physics and Margenau, 1984

Then came in 1984 the book by the distinguished quantum physicist Henry Margenau *The Miracle of Existence*. It was a light at the end of the tunnel to read:

> The mind may be regarded as a field in the accepted physical sense of the term, but it is a non-material field, its closest analogue is perhaps a probability field . . . nor is it required to contain energy in order to account for all the known phenomena in which mind interacts with the brain.

I immediately communicated with Margenau, whom I already knew, and we had a stimulating correspondence. Meanwhile I had written a most appreciative review of his book for *Foundations of Physics*.

It was just at this time that Professor Herman Haken of Stuttgart invited me to a conference that he was organizing for 6–11 May 1985, at the attractive establishment at Schloss Elmau in Southern Bavaria. The conference was on the general theme of 'Complex Systems: Operational Approaches in Neurobiology, Physics and Computers'. It gave me the opportunity to present my paper 'New Light on the Mind–Brain Problem: How Mental Events Could Influence Neural Events'.

Professor Haken was most encouraging that I should give a full presentation of the new developments, so I was given two hours on the programme for lecture and discussion, but even that was not enough. The discussion was resumed for an hour or so after dinner! I was very happy at the good reception given by the group of about 60 physical scientists, computer theorists, and other scientists. My paper was published in the conference volume by Springer-Verlag, and appropriately I have included it as the first of my published papers in this book (Chapter 4).

References to Chapters 1 and 2

Armstrong, D. M. (1981) *The Nature of Mind* (Cornell University Press, Ithaca NY).
Barlow, H. B. (1972) Single units and sensation: A neuron doctrine for perceptual psychology?, *Perception* **1**, 371–394.
Beck, F., and Eccles, J. C. (1992) Quantum aspects of consciousness and the role of consciousness, *Proc. Nat. Acad. Sci.* **89**, 11357–11361.
Beloff, J. (1962) *The Existence of Mind* (Mecgibbon and Kee, London).
Birch, C. (1974) Chance, necessity and purpose, in *Studies in the Philosophy of Biology*, edited by F. J. Ayala and T. Dobzhansky (Macmillan, London), pp. 225–239.
Blakemore, C. (1977) *Mechanics of the Mind* (Cambridge University Press, Cambridge).
Bunge, M. (1980) *The Mind–Body Problem* (Pergamon, Oxford).
Changeux, J. P. (1985) *Neuronal Man. The Biology of Mind* (Pantheon, New York).
Creutzfeldt, O. D., and Rager, G. (1978) Brain mechanisms and the phenomenology of conscious experience, in *Cerebral Correlates of Conscious Experience*, edited by P. A. Buser and A. Rougeul-Buser (North Holland, Amsterdam), pp. 311–318.
Deecke, L., and Kornhuber, H. H. (1978) An electrical sign of participation of the mesial 'supplementary' motor cortex in human voluntary finger movement, *Brain Res.* **159**, 473–476.
Dobzhansky, T. (1967) *The Biology of Ultimate Concern* (New American Library, New York).
Doty, R. W. (1975) Consciousness from neuron, *Acta Neurobiol. Exp. (Warsz.)* **35**, 791–804.
Eccles, J. C. (1951) Hypotheses relating to the brain–mind problem, *Nature* **168**, 53–57.
Eccles, J. C. (1953) *The Neurophysiological Basis of Mind: The Principles of Neurophysiology* (Clarendon, Oxford).
Eccles, J. C. (1966) in *Brain and Conscious Experience*, Symposium of the Pontif. Acad. of Sci. (Springer, Berlin, Heidelberg).
Eccles, J. C. (1970) *Facing Reality* (Springer, Berlin, Heidelberg).
Eccles, J. C. (1979) *The Human Mystery* (Springer, Berlin, Heidelberg).
Eccles, J. C. (1980) *The Human Psyche* (Springer, Berlin, Heidelberg).
Eddington, A. S. (1939) *The Philosophy of Physical Science* (Cambridge University Press, Cambridge).
Edelman, G. M. (1978) Group selection and phasic reentrant signalling: A theory of higher brain function, in *The Mindful Brain*, edited by F. O. Schmitt (MIT Press, Cambridge MA).
Edelman, G. M. (1989) *The Remembered Present. A Biological Theory of Consciousness* (Basic Books, New York).
Feigl, H. (1967) *The Mental and the Physical* (University of Minnesota Press, Minneapolis).
Granit, R. (1977) *The Purposive Brain* (MIT Press, Cambridge MA).
Hinshelwood, C. (1962) *The Vision of Nature* (Cambridge University Press, Cambridge).
Ingvar, D. H. (1975) Patterns of brain activity revealed by measurements of regional cerebral blood flow, in *Brain Work*, edited by D. H. Ingvar and N. A. Lassen (Munksgaard, Copenhagen).

Ingvar, D. H. (1990) On ideation and 'ideography', in *The Principles of Design and Operation of the Brain*, edited by J. C. Eccles and O. Creutzfeldt (*Experimental Brain Research*, Series 21) (Springer, Berlin, Heidelberg), pp. 433–453.

Kneale, W. (1962) *On Having a Mind* (Cambridge University Press, Cambridge).

Kornhuber, H. H. (1978) A reconsideration of the brain–mind problem, in *Cerebral Correlates of Conscious Experience*, edited by P. A. Buser and A. Rougeul-Buser (North Holland, Amsterdam), pp. 319–331.

Laszlo, E. (1972) *Introduction to Systems Philosophy* (Gordon and Breach, New York).

Lorenz, K. (1977) *Behind the Mirror* (Methuen, London).

MacKay, D. M. (1978) Selves amd brains, *Neurosciences* **3**, 599–606.

Margenau, H. (1984) *The Miracle of Existence* (Ox Bow, Woodbridge CT).

Monod, J. (1971) *Chance and Necessity* (Knoff, New York).

Mountcastle, V. B. (1978) An organizing principle for cerebral function: The unit module and the distributed system, in *The Mindful Brain*, edited by F. O. Schmitt (MIT Press, Cambridge MA).

Penfield, W. (1975) *The Mystery of the Mind* (Princeton University Press, Princeton NJ).

Polanyi, M. and Prosch, H. (1975) *Meaning* (The University of Chicago Press, Chicago).

Polten, E. (1975) *Critique of the Psycho-physical Identity Theory* (Mouton, Paris).

Popper, K. R. (1959) *The Logic of Scientific Discovery* (Hutchinson, London).

Popper, K. R. (1968) On the theory of the objective mind, in *Akten des XIV Internationalen Kongresses für Philosophie*, Vol. 1 (Herder, Vienna), pp. 25–53.

Popper, K. R. (1972) *Objective Knowledge: An Evolutionary Approach* (Clarendon, Oxford).

Popper, K. R., and Eccles, J. C. (1977) *The Self and Its Brain* (Springer, Berlin, Heidelberg).

Pribram, K. H. (1971) *Languages of the Brain* (Prentice-Hall, Englewood Cliffs NJ).

Ramón y Cajal, S. R. (1911) *Histologie du Système Nerveux de l'Homme et des Vertébrés*, Vol. II (Maloine, Paris).

Rensch, B. (1971) *Biophilosophy* (Columbia University Press, New York).

Rensch, B. (1974) Polynomistic determination of biological processes, in *Studies in the Philosophy of Biology*, edited by F. J. Ayala and T. Dobzhansky (Macmillan, London), pp. 241–258.

Schrödinger, E. (1958) *Mind and Matter* (Cambridge University Press, Cambridge).

Searle, J. R. (1984) *Minds, Brains and Science* (British Broadcasting Corporation, London).

Searle, J. R. (1992) *The Rediscovery of the Mind* (MIT Press, Cambridge MA).

Shapere, D. (1974) Discussion of Rensch, in *Studies in the Philosophy of Biology*, edited by F. J. Ayala and T. Dobzhansky (Macmillan, London).

Sherrington, C. S. (1940) *Man on His Nature* (Cambridge University Press, London). (Second edition 1951.)

Smart, J. J. C. (1963) *Philosophy and Scientific Realism* (Routledge and Kegan Paul, London).

Sperry, R. W. (1977) Forebrain commissurotomy and conscious awareness, *J. Med. Phil.* **2**, 101–126.

Sperry, R. W. (1980) Mind–brain interaction: Mentalism, yes; dualism, no, *Neuroscience* **5**, 195–206.

Szentágothai, J. (1975) The 'module' concept in cerebral cortex architecture, *Brain Res.* **95**, 475–496.

Szentágothai, J. (1978) The neuron network of the cerebral cortex. A functional interpretation, *Proc. Roy. Soc. London B* **201**, 219–248.

Szentágothai, J. (1979) Local neuron circuits of the neocortex, in *The Neurosciences. Fourth Study Program*, edited by F. O. Schmitt and F. G. Warden (MIT Press, Cambridge MA).

Thorpe, W. H. (1974) *Animal Nature and Human Nature* (Methuen, London).

Thorpe, W. H. (1978) *Purpose in a World of Chance* (Oxford University Press, Oxford).

Waddington, C. H. (1969) The theory of evolution today, in *Beyond Reductionism*, edited by A. Koestler and J. R. Smythies (Hutchinson, London).

Wigner, E. P. (1964) Two kinds of reality, *Monist* **48**, 248–264.

3 Recent Theoretical Studies on the Mind–Brain Problem

In recent years the literature on the mind–brain problem has become enormous and diverse. Any attempt at quotation of all important contributions would produce disorganized chapters. So I have chosen to present in this chapter some critical abstracts of the literature published in many books by leading authorities.

For ease of reference the sections are arranged alphabetically by author. Authors are selected for discussion because their publications relate specifically to the chapters of this book. Further reference to this literature will be made throughout the book.

3.1 J. P. Changeux, *Neuronal Man: The Biology of Mind*, 1985

Much of Changeux's book is written for a popular audience. It is not my task to criticize any shortcomings that there are in this ambitious attempt to present the science of the brain. I will concentrate on those concepts in the book that relate to the assembled Chapters 4 to 9 of the present work and provide background.

The centre of Changeux's neurophilosophy is given in the long Chapter 5 on 'Mental Objects'.

On p. 127 enters Neuronal Man. I quote:

> The human brain can develop strategies on its own. It anticipates coming events and elaborates its own programs. This capacity for self is one of the most remarkable features of the human cerebral machine, and its supreme product is thought.

This appears to be a dualism that I would not reject at first, but then it appears as a subtle expression of the identity theory of Feigl (Chapter 1.3.4 of the present book), with the brain being given a 'capacity for self'. On p. 126 Changeux states that

> a computer possesses properties that are radically different from those of the cerebral machine.

I also agree.

The mental object is identified in several specific cortical areas. Then there follows for over two pages (pp. 138–140) a neuronal fantasia with no mention of how a 'mental object' becomes mental.

In 'Problems of Consciousness' Changeux writes:

> While we are awake and attentive, we appreciate and pursue the formation of precepts and concepts. We can store and recall mental objects, link them together, and recognize their resonance. We are conscious of all this in our unending dialogue with the outside world, but also within our own inner world, our 'me'. (p. 145)

When considering the role of the reticular activating system that was shown by Moruzzi and Magoun to provide the essential background activity of the cerebral cortex, Changeux states:

> Consciousness, then, corresponds to a regulation of the overall activity of cortical neurons and, more generally, of the entire brain. A few small groups of neurons in the brainstem, with their cell bodies centrally situated, exert a 'global' influence, thanks to the divergent nature of their axons. (p. 151)

As Changeux expresses it,

> the different groups of neurons in the *reticular formation* inform each other of their mutual activity. They form a system of hierarchical, parallel pathways in permanent *reciprocal contact* with the other structures of the brain. A holistic *integration* between various centers results. From the interplay of these linked regulatory systems, consciousness is born. (p. 158)

I give these extensive extracts to show this extraordinary explanation of the origin of consciousness by complex neuronal network interaction, which resembles in general the consciousness model given by Edelman. It is entirely based on neuronal network interaction according to classical physics, and hence according to Stapp it will not do.

In 'The Substance of the Spirit' Changeux states:

> The theme is to destroy the barriers that separate the neural from the mental and construct a bridge, however fragile, allowing us to cross from one to the other. (p. 168)

But is this not the program of dualist-interactionism? Changeux seems to have missed the way that was earlier described but is later developed fully in Chapters 4–9 of the present book.

This 'mental objects' chapter finishes with much speculation on the neuronal composition of mental objects, even with 'homuncular' components!

> The combinational possibilities provided by the number and diversity of connections in the human brain seem quite sufficient to account for human capabilities. There is no justification for a split between mental and neuronal activity. (p. 275)

> It seems quite legitimate to consider that mental states and physiological or physico-chemical states of the brain are identical. (p. 275)

So Changeux lapses into a simple identity theory without any more philosophical ado.

> We must first construct within our brain an image of 'man', an idea, like a model, that we can contemplate. (p. 284)

> Man has no longer need of a Spirit. It is enough for him to be 'Neuronal Man'. (p. 169)

Why not also 'Mindful Woman'?

3.2 F. Crick and C. Koch, 'Towards a Neurobiological Theory of Consciousness', 1990

Implicit in the concepts of Crick and Koch is the identity theory of Feigl that has been described in Chapter 1.3.4 and Fig. 1.2. They state:

> Our basic hypothesis at the neural level is that it is useful to think of consciousness as being correlated with a special type of activity of perhaps a subset of neurons in the cortical system . . . At any moment consciousness corresponds to a particular type of activity in a transient set of neurons that are a subset of a much larger set of potential candidates. (p. 266)

> Our basic idea is that consciousness depends crucially on some form of rather short-term memory and also on some form of serial attentional mechanism. This attentional mechanism helps sets of the relevant neurons to fire in a coherent semioscillatory way, probably at a frequency in the 40–70 Hz range, so that a temporary global unity is imposed on neurons in many different parts of the brain. These oscillations then activate short-term (working) memory. (p. 263)

But there is much evidence that mental attention can cause neural activation, as has been described by Beck and myself (Beck and Eccles 1992 and Chapters 9 and 10.6 of the present book). Crick and Koch can now use these new discoveries to account for the choice of the most salient objects for attention with activation of the most appropriate neurons to give

[the] best interpretation of which we become aware.

> The information about a single object is distributed about the brain. There has, therefore, to be a way of imposing a temporary unity on the activities of all the neurons that are relevant at that moment. (Incidentally we see no reason at all why this global unity should require fancy quantum effects.) The achievement of this unity may be assisted by a fast attentional mechanism, the exact nature of which is not yet understood. (p. 274)

So far so good, but their sequel cannot be accepted. It concerns

> machines having complex rapidly changing and highly parallel activity. When we can both construct such machines and understand their detailed behavior much of the mystery of consciousness may disappear. (p. 274)

This can be dismissed as science fiction of a blatant kind. Reference should be made to the sections in this chapter on Penrose (3.7) and Searle (3.8 and 3.9) and to Chapter 10.7 for a criticism of such artificial intelligence machines.

3.3 F. Crick and C. Koch, 'Consciousness', *Scientific American*, 1992

After an introductory discussion largely covered by the preceding paper, Crick and Koch state:

> There may be a very transient form of fleeting awareness that represents only rather simple features and does not require an attentional mechanism. From this brief awareness the brain constructs a viewer-centered representation—what we see vividly and clearly—that does require attention. This in turn probably leads to three-dimensional object representations and thence to more cognitive ones. (p. 112)

I think that these are important statements suggesting four levels of consciousness. I also agree with their closing statement:

> Once we have mastered the secret of this simple form of awareness, we may be close to understanding a central mystery of human life: how the physical events occurring in our brains while we think and act in the world relate to our subjective sensations—that is, how the brain relates to the mind. (p. 117)

There appears to be no insistence on the identity theory and no denial of dualism. The way is open to the new findings and hypotheses of Beck and myself (Beck and Eccles 1992 and Chapter 9 of the present book).

3.4 D. C. Dennett, *Consciousness Explained*, 1991

Has Dennett dunnit?

> Human consciousness is just about the last surviving mystery. (p. 21)

It is still a mystery at the end of his 468-page book *Consciousness Explained*!

To start with I quote from his initial summary:

> We have found four reasons for believing in mind stuff. The conscious mind, it seems, cannot just be the brain, or any proper part of it, because nothing in the brain could
> 1 be the medium in which the purple cow is rendered;
> 2 be the thinking thing, the *I* in 'I think, therefore I am';
> 3 appreciate wine, hate racism, love someone, be a source of *mattering*;
> 4 act with moral responsibility.
>
> An acceptable theory of human consciousness must account for these four compelling grounds for thinking that there must be mind stuff. (pp. 32–33)

This is dualism: brain and mind.

Yet this section leads straight on to the section headed 'Why Dualism is Forlorn':

> The prevailing wisdom, variously expressed and argued for, is *materialism*: there is only one sort of stuff, namely *matter*—the physical stuff of physics, chemistry, and physiology—and the mind is somehow nothing but a physical phenomenon. In short, the mind is the brain. (p. 33)

> Let us concentrate on the returned signals, the directives from mind to brain. These, *ex hypothesi*, are not physical; they are not light waves or sound waves or cosmic rays or streams of subatomic particles. No physical energy or mass is associated with them. How, then, do they get to make a difference to what happens in the brain cells they must affect, if the mind is to have any influence over the body? A fundamental principle of physics is that any change in the trajectory of any physical entity is an acceleration requiring the expenditure of energy, and where is this energy to come from? It is this principle of the conservation of energy that accounts for the physical impossibility of 'perpetual motion machines', and the same principle is apparently violated by dualism. This confrontation between quite standard physics and dualism has been endlessly discussed since Descartes's own day, and is widely regarded as the inescapable and fatal flaw of dualism. (p. 35)

But now in Chapter 9 of the present book Beck and I establish that there is no such flaw, so dualism is not forlorn any longer!

Dennett makes an interesting comment on p. 39:

> Some brain researchers today would never dream of mentioning the mind or anything 'mental' in the course of their professional duties. For other, more theoretically daring researchers, there is a new object of study, the mind/brain. This newly popular coinage nicely expresses the prevailing materialism of these researchers, who happily admit to the world and themselves that what makes the brain particularly fascinating and baffling is that somehow or other it is the mind. But even among these researchers there is a reluctance to confront the Big Issues, a desire to postpone until some later date the embarrassing questions about the nature of consciousness. (p. 39)

Dennett is here expressing what I was writing in Chapter 1.

In Chapter 8 Dennett expresses his philosophy:

> In our brains there is a cobbled-together collection of specialist brain circuits, which, thanks to a family of habits inculcated partly by culture and partly by individual self-exploration, conspire together to produce a more or less orderly, more or less effective, more or less well-designed virtual machine, *the Joycean machine.* By yoking these independently evolved specialist organs together in common cause, and thereby giving their union vastly enhanced powers, this virtual machine, this software of the brain, performs a sort of internal political miracle. It creates a *virtual captain* of the crew, without elevating any one of them to long-term dictatorial power. (p. 228)
>
> The resulting executive wisdom is just one of the powers traditionally assigned to the Self, but it is an important one. (p. 228)

Dennett gives this brief initial sketch of his theory:

> There is no single, definitive 'stream of consciousness', because there is no central Headquarters, no Cartesian Theater where 'it all comes together' for the perusal of a Central Meaner. Instead of such a single stream (however wide), there are multiple channels in which specialist circuits try, in parallel pandemoniums, to do their various things, creating Multiple Drafts as they go. Most of these fragmentary drafts of 'narrative' play short-lived roles in the modulation of current activity but some get promoted to further functional roles, in swift succession, by the activity of a virtual machine in the brain. (p. 254)
>
> They are often opportunistically enlisted in new roles, for which their native talents more or less suit them. The result is not bedlam only because the trends that are imposed on all this activity are themselves the product of design. Some of this design is innate, and is shared with other animals. But it is augmented, and sometimes even overwhelmed in importance, by micro-habits of thought that are developed in the individual, partly idiosyncratic results of self-exploration and partly the predesigned gifts of culture. Thousands of memes, mostly borne by Language, but also by wordless 'images' and other data structures, take up residence in an individual brain, shaping its tendencies and thereby turning it into a mind. (p. 254)

This series of dogmatic statements is the beginning of Dennett's Multiple Drafts Theory. I ask, is this a theory of consciousness? Dennett apparently realizes this when he states:

> I have been coy about consciousness up to now. Until the whole theory-sketch was assembled, I had to deflect such doubts, but at last it is time to grasp the nettle, and confront consciousness itself, the whole marvellous mystery. And so I hereby declare that YES, my theory is a theory of consciousness. Anyone or anything that has such a virtual machine as its control system is conscious in the fullest sense, and is conscious *because* it has such a virtual machine. (p. 281)

One hundred and seventy pages later Dennett states:

> I haven't replaced a metaphorical theory, the Cartesian Theater, with a non-metaphorical ('literal, scientific') theory. All I have done, really, is to replace one family of metaphors and images with another, trading in the Theater, the Witness, the Central Meaner, the Figment, for Software, Virtual Machines, Multiple Drafts, a Pandemonium of Homunculi. It's just a war of metaphors, you say—but metaphors are not 'just' metaphors; metaphors are the tools of thought. (p. 455)

I am much dissatisfied by Dennett's dogmatic creation, the Multiple Drafts Model, because it discounts a unique Self which is central to our experience. Dennett wants to get rid of the Cartesian Theater, but all he seems to finish with is emptiness. I miss the enchantment, the wonder, the challenge that is experienced by my unique Self. Why does Dennett perpetrate this impoverished and empty theory on us. Is he lacking in humility?

(For an extensive review, at times critical, see Roskies and Wood, *The Sciences*, May/June 1992, p. 44.)

3.5 G. M. Edelman, *The Remembered Present*, 1989

Let me first emphasize that I am not presenting a review of this monumental book. I would not dare! My task is to try to describe Edelman's efforts to give a solution of the mind–brain problem. He concludes that it is at least feasible to construct a detailed and principled brain-based theory of consciousness. Edelman starts out with a stark challenge:

> Any adequate global theory of brain function must include a scientific model of consciousness, but to be scientifically acceptable it also must avoid the Cartesian dilemma. In other words, it must be uncompromisingly physical. (p. 10)

I should read no further! This is the solution of reductionist materialism, which he calls 'materialist metaphysics'. He goes on to state:

> Scientific epistemology must confront the issue of consciousness in terms of evolution, development, brain structure, and the physical order as we know it. If the confrontation is to remain in the scientific domain, a dualistic solution or any form of Cartesian empiricism cannot be countenanced, [being] often accompanied by what might be called Cartesian shame. (p. 278)
>
> An example lacking any Cartesian shame is seen in K. R. Popper and J. C. Eccles's *The Self and Its Brain*, which is an explicit dualist account. (p. 278, Footnote 25)

This prevailing critical atmosphere bleakly recalls an inverse Inquisitorial degradation!

There is no detailed structural or functional consideration. Just network theories with identity theory will give consciousness. Edelman writes a section entitled

> The insufficiency of functionalism with its basis on computational states of the brain. Brains and consciousness are not based on Turing machines.

Edelman is devoted to his Theory of Neuronal Group Selection (TNGS), which is no more than an elaborate example of a network theory. However, through the book TNGS becomes almost personalized with its own particular wisdom.

When correlating 'brain' states with 'mental' states in the identity hypothesis Edelman writes:

> Many different brain states can lead to a single, particular conscious state. There is a token identity in the sense that a mental event is a physical event, although as a process it has properties that cannot be identical to those of the structural components of the brain that give rise to it. (p. 260)

I would warn against taking this identity too strictly: it is philosophese not physics!

When 'The Model of Primary Consciousness' is finally unveiled on pp. 153–155 it appears as a very sophisticated type of identity theory derived from a great complexity of neural networks, described on pp. 151–153, with reentrant interaction and elaborated in far more detail than in any previous account, e.g. by Feigl (1957). Finally Edelman claims that

> this reentrant interaction between a special form of memory with strong conceptual components (we shall consider its possible loci later) and a stream of perceptual categorizations would generate primary consciousness. (p. 151)

This seems to me like a conjuring trick! The subsequent account gives the proposed neural pathways. Great stress is placed on memory.

Edelman has been devoted to neural circuits (the TNGS) that form the identity with the experiences of primary consciousness. There is a fantastic elaboration that could relate the phenomenal experiences such as raw feels in all their immense variety of qualia. However, Edelman claims that

> as scientists, we can have no concern with ontological mysteries concerned with *why* there is something and not nothing, or *why* warm feels 'warm'. (p. 168)

I quote a valuable introduction to 'Higher-Order Consciousness':

> Speech allows the development of internal models and conceptual categorizations that can be time-independent as well as of enriched distinctions of the concept of self, leading to personhood. (p. 185)

Evidently we are 'unknowingly' entering dualism: where else is the self or the person? Unfortunately Edelman has been misled by Lieberman (1975) in *Hominid Evolution* to give an absurdly late development of language in relation to brain development, which came as early as *Homo habilis* (Tobias 1983; Stebbins 1982).

> To become conscious of consciousness, or directly aware, an animal (human) must have a distinction of self from nonself that is to some degree time-independent. (p. 186)

> At this point, 'self' becomes a term referring to such a conceptual model and not just to a biological individual, as is the case for primary consciousness. (p. 187)

Thus we are introduced to personhood. We are far from the materialist metaphysics proclaimed in Edelman's introduction!

> Subjective features relating to self-awareness, first-person usage, meaning, and the like are just that—subjective processes going on in a person who has higher-order consciousness. (p. 194)

> The causal relations between physical and mental events can be understood in terms of the heterogeneity of consciousness arising as a special form of memory connected with the self–nonself distinctions that are related by reentry to perceived events. (p. 194)

This is dualist-interactionism.

> Volition requires the awareness of a goal and the ability to direct action (and thus sensation and perception). (p. 197)

> It is on these imperfect means and on the rich affective exchange based on an inner life that personhood is built. (p. 251)

All these and other quotations present expressions that I would be happy to make as a dualist! There is need for humility!

Edelman claims to have outlined a biological theory of consciousness that is linked to the structure and physiology of the brain. However, this neuroscience is at a relatively crude level and completely overlooks the ultimate refinements that are fundamental to the mind–brain theory, as described in Chapters 4–9 of the present book.

> Extreme reductionist positions that expect *to account for* consciousness on the basis of quantum mechanics and that ignore the facts of evolution seem overambitious and empty (Margenau 1984; Wigner 1979). (p. 254)

It is embarrassing for Edelman to display his ignorance by accusing these both of extreme reductionist positions. On the contrary, they opened the way for dualist-interactionism by quantum physics (Chapters 4–9 of the present book).

Edelman has a 'conviction' (p. 261) that he has some degree of free will but cannot account for it by coming to a compatibilism like Searle for a conscious rational agent.

What I find missing is the recognition of the unity of the self and its uniqueness. There is a wide disparity in different aspects of Edelman's philosophy, and he does it all with classical physics.

3.6 D. Hodgson, *The Mind Matters*, 1991

It has to be recognized that Judge Hodgson brings a highly skilled legal mind to bear on the problem of the human mind with respect to such procedures as volition, freedom, responsibility, and open choice.

Concerning *mental events*, one can accept the initial statement that mental events are the fully conscious experiences of normal human beings.

Concerning *perception*, I will deal, in Chapter 10.5, with the problems under the heading of attention by which the mind selects and intensifies the experiences in a continuous series of reappraisals. Hodgson rather neglects the great experiences of qualia. With respect to Cartesian dualism with two kinds of substance, I think it better to use the concepts of Popper (Figure 1.1), Worlds 1 and 2, and not to use the double aspect theory.

I shall not engage here in a detailed appraisal of the consensus views by the wide range of authorities in Hodgson's Chapter 3. With the new insights provided by Beck and myself in Chapter 9 of the present book it is no longer profitable to pursue all the detailed ramifications of the consenses views which Hodgson has assembled like a comprehensive legal document dealing with philosophy, computer science, brain science, and quantum physics. I can only speak critically on the brain science, which should be central to a book entitled *The Mind Matters*, but which on pp. 83–90 is inadequate.

Hodgson considers briefly my ideas as expressed in a very out-of-date book (Eccles and Popper 1977). He would have been greatly helped by my two Gifford Lectures (1977, 1978), which should have been available in Sydney. It is better if I move on from pp. 83–90, because a greatly different account of mental events and the mind–brain problem is given in Chapters 5–10 of the present book.

It is unfortunate that Hodgson has dedicated so much effort to attaining such a good general understanding of quantum physics, which occupies Part III of his book (180 pages). Undoubtedly quantum physics is of great significance, as was already appreciated by Margenau (1984), and fully exploited by Beck and myself (Beck and Eccles 1992 and Chapter 9 of the present book) in relation to special microsites of the neocortex. I have critically considered artificial intelligence and robotics in Chapter 10.6 and nothing more need be said when considering the consensus.

Concerning *consciousness selected by evolution* Hodgson appreciates very much the importance of consciousness for survival. It could lead to understanding with more skill in communal existence and cooperative learning. An attempted explanation of how consciousness came about is given in Chapter 7 of the present book. With higher levels of evolution to the eventual emergence of *Homo sapiens sapiens* self-consciousness could give wisdom in judgement in survival situations with freedom from mechanistic determinism and the ability to reason informally. This is our ubiquitous practice when not constrained by philosophers!

> We can know what the conscious experiences of others are like only because we have conscious experience ourselves. It is a smallish step from this to asserting that we can *understand* the actions of others only because we can know the conscious experiences underlying and motivating them, and thus only because we ourselves have such conscious experiences.

Hodgson writes well on

> the central role which mind has in the life and world of every human being: in an important sense it is the totality of each person's world, it is basic to all knowledge, and may well be the constructor (not merely the knower) of the common external reality as we know it.

Mind is the totality, but we have to recognize that it is much more extensive and richer than what we seem to recognize as the totality of our world. It is certain that I can go on writing about my remembered world for many hundreds of pages. As I write, long 'forgotten' aspects of my life are reexperienced. There seems to be no end to this discovery of the experiential richness when I am writing my autobiography with minimal use of documents.

> Consensus writers, when they do consider consciousness at all, tend to treat it as a small anomaly or embarrassment in an otherwise perfect mechanistic physical world; whereas I say that consciousness is of central importance—without it, there could be no point to anything. (p. 447)

Much that Hodgson writes on personal identity and the self is just repeating the science fiction of Parfit (pp. 409–413) with brain transplants and reversible blockage of the corpus callosum etc. Parfit is a reductionist.

In sketches for a world view by Hodgson there is an account of Descartes's dualism of substance that I join Hodgson in rejecting. The concept of substance leads to a materialist aspect of the mind. I speak instead of the spiritual existence of the self without mentioning any 'substance' properties. The great problem is 'how the self controls its brain'. This is dualistic, but not in terms of two substances. Instead it relates to the two worlds of Popper.

As Hodgson develops his philosophy in Chapter 19, it is rather close to my expression of dualism with even a relevant conscious entity in the key role, which I would call the self:

> Although the dualism of this book does not depend upon the existence of two different substances, the distinction drawn between the physical and the mental is of sufficient moment to make it reasonable to follow Popper's terminology, and to identify two 'worlds': the World 1 of the physical or objective and World 2 of the mental or subjective. (p. 445)

On p. 449 morality is discussed in a good judgemental account with which I approve, except that 'love' is not mentioned!

On p. 453 Hodgson goes on to consider the soul that goes beyond the self in that it is not matter dependent and so may exist without the brain. He speculates on survival after death in relation to the essence of a person, which is more than I would dare! We have to be very humble in our efforts to understand these profound mysteries. I respect Hodgson's enquiry into the essential psyche before conception and after death. However, I don't find the divisiveness or the other speculations attractive. I shelter in humility; I agree with Hodgson that religious beliefs should be regarded as an approximation to the inexpressible truth in an allegorical representation of our belief. I do hope for a unique self or soul for survival.

I will not discuss the next section on God because it is not relevant to the theme of this book, but I approve in general.

In his last section, on the purpose of life, I respect Hodgson's dedicated quest for truth and values, but he should have devoted more time to the brain, which amazingly is so neglected by the materialists, who are intellectually very lazy. I hope that Hodgson will in the future give the study of the brain even a fraction of the time he has devoted to quantum physics. The way that Beck and I have opened it up is illimitable, but materialist philosophers will ignore this, as will the whole assemblage Hodgson lumps together as the consensus!

As a final quotation Hodgson writes well:

> What we are looking for, I think, is a purpose that we should try to recognize and pursue, which our lives if correctly lived will fulfil in fact, and which is right and good. It may also be God's purpose for us. I think that the notion of a purpose of life makes more sense in relation to a God, or some wider consciousness, than it otherwise would. (p. 462)

3.7 R. Penrose, *The Emperor's New Mind*, 1991

This book is a sophisticated criticism of the extravagant claims made for artificial intelligence (AI) machines.

The first chapter is 'Can a Computer Have a Mind?' The Prologue and Epilogue give an entertaining vision of the grand opening ceremony of the Ultronic Computer, which has unbelievable intelligence and memory but is unable to appreciate and answer the only question it was asked, a child's simple question, 'What does it feel like?'.

I am reminded of an episode at the great opening ceremony of the grandiose scientific laboratories of Western Reserve University of Cleveland in 1962. I was the last speaker, being given the attractive title 'Mind: the ultimate expression of the living state'. In the discussion, presided over by Professor Philip Abelson and carried on by him, I was presented with the unbelievable future of artificial intelligence machines as against my lecture on the wonder of evolved brains with minds. It was not just artificial intelligence, as in chess games, but a performance in understanding and creating and feeling and meaning and remembering that would far surpass humans, so he said! In desperation at this prolonged and emotional confrontation I finally shouted out: 'Would you like your daughter to marry one?'. The 5000 students screamed with delight, but I have never been forgiven!

My commentary can begin with Penrose's question on p. 402:

> Is our picture of a world governed by the rules of classical and quantum theory, as these rules are presently understood, really adequate for the description of brains and minds?

My reaction is that we have to go on scientifically and philosophically, and we can be greatly encouraged by the progress. There of course are many blind alleys that have enormous attraction to computer technologists, notably the artificial intelligence and robotics machines. I agree with Penrose's general rejection of such models of intelligence and consciousness. A related project is to study the properties of assumed neuronal networks, which can be modelled by computer technology, and that even may give an opening to robotics.

Penrose asks two challenging questions:

> How is it that a material object (a brain) can actually *evoke* consciousness? (p. 405)

and, conversely:

> How is it that a consciousness, by the action of its will, actually *influences* the (apparently physically determined) motion of material objects? (p. 405)
>
> These are the passive and active aspects of the mind–body problem. It appears that we have in 'mind' (or, rather, in 'consciousness'), a non-material 'thing' that, on the one hand, is evoked by the material world and, on the other, can influence it. (p. 405)

It seems that Penrose is satisfied with an explanation based on the identity theory.

Penrose tries to explore the nature of his conscious experiences:

> To be conscious, I seem to have to be conscious *of* something, perhaps a sensation such as pain or warmth or a colourful scene or a musical sound; or perhaps I am conscious of a feeling such as puzzlement, despair, or happiness; or I may be conscious of the memory of some past experience, or of coming to an understanding of what someone else is saying, or a new idea of my own; or I may be consciously intending to speak or to take some other action such as get up from my seat. I may also 'step back' and be conscious *of* such intentions or of my feeling of pain or of my experiencing of a memory or of my coming to an understanding; or I may even just be conscious of my own consciousness. I may be asleep and still be conscious to some degree, provided that I am experiencing some dream; or perhaps, as I am beginning to awake, I am consciously influencing the direction of that dream. I am prepared to believe that consciousness is a matter of degree and not simply something that is either there or not there. (p. 407)

Penrose asks the important question:

> What does consciousness actually do? . . . What is it that we can *do* with conscious thought that cannot be done unconsciously? . . . Somehow, consciousness is needed in order to handle situations where we have to form new judgements, and where the rules have not been laid down beforehand.

> It is hard to be very precise about the distinction between the kinds of mental activity that seem to require consciousness and those that do not. (p. 411)

> *The judgement-forming* that I am claiming is the hallmark of consciousness is *itself* something that the AI people would have no concept of how to program on a computer. (p. 412)

> I am putting forward the argument here that it is this ability to divine (or 'intuit') truth from falsity (and beauty from ugliness!), in appropriate circumstances, that is the hallmark of consciousness. (p. 412)

Penrose expresses well the way our wise performance is essentially based on consciousness, stating:

> I am referring to the judgements that one continually makes while one is in a conscious state, bringing together all the facts, sense impressions, remembered experiences that are of relevance, and weighing things against one another—even forming inspired judgements, on occasion. Enough information is in principle available for the relevant judgement to be made, but the process of formulating the appropriate judgement, by extracting what is needed from the morass of data, may be something for which no clear algorithmic process exists. (p. 412)

> I am guessing that consciousness would, under such circumstances, come into its own as a means of conjuring up the appropriate judgements. (p. 413)

After much discussion Penrose comes to a wise conclusion:

> I cannot believe that the anthropic argument is the *real* reason (or the only reason) for the evolution of consciousness. There is enough evidence from other directions to convince me that consciousness *is* of powerful selective advantage, and I do not think that the anthropic argument is needed. (p. 434)

In 'Time Delays of Consciousness' (p. 439) Penrose is quite rightly concerned with some of the findings of these searching investigations, but in part they involve technical problems, as yet unresolved, so they should not be discussed here. Reference could be made to Libet (1990, p. 185 and the General Discussion on pp. 207–209).

Penrose is of great interest because he concentrates at the end of his book on the great problems:

> Some readers may, from the start, have regarded the 'strong-AI-supporter' as perhaps largely a straw man! Is it not 'obvious' that mere computation cannot evoke pleasure or pain; that it cannot perceive poetry or the beauty of an evening sky or the magic of sounds; that it cannot hope or love or despair; that it cannot have a genuine autonomous purpose? Yet science seems to have driven us to accept that we are all merely small parts of a world governed in full detail (even if perhaps ultimately just probabilistically) by very precise mathematical laws. Our brains themselves, which seem to control all our actions, are also ruled by these same precise laws. The picture

has emerged that all this precise physical activity is, in effect, nothing more than the acting out of some vast (perhaps probabilistic) computation—and, hence, our brains and our minds are to be understood solely in terms of such computations . . . Yet *it is hard to avoid an uncomfortable feeling that there must always be something missing from such a picture.*

This is the picture of monist materialism that I have been contending with and rejecting all my life. Penrose goes on to state:

Consciousness seems to me to be such an important phenomenon that I simply cannot believe that it is something just 'accidentally' conjured up by a complicated computation. It is the phenomenon whereby the universe's very existence is made known. One can argue that a universe governed by laws that do not allow consciousness is *no universe at all.* I would even say that all the mathematical descriptions of a universe that have been given so far must fail this criterion. It is only the phenomenon of consciousness that can conjure a putative 'theoretical' universe into actual existence!' (p. 447)

My comment is that for each of us consciousness gives the experience in all of its infinite variety of a unique self or soul. Surely this is the answer that Penrose is seeking for and that he puts aside when he says:

We shall be having enough trouble with coming to terms with 'consciousness' as it stands, so I hope that the reader will forgive me if I leave the further problems of 'mind' and 'soul' essentially alone! (p. 407)

3.8 J. R. Searle, *Minds, Brains and Science*, 1982

Searle initially writes a simple philosophical story that I find attractive. He states that there is a temptation to downgrade the status of mental entities. So most of the recently fashionable materialist conceptions of the mind—such as behaviourism, functionalism, and physicalism—end up by denying, implicitly or explicitly, that there are any such things as minds as we ordinarily think of them. That is, they deny that we do really *intrinsically have* subjective, conscious, mental states that are as real and as irreducible as anything else in the universe.

This denial prompts Searle to counterattack by listing four kinds of mental phenomena that cannot be fitted into scientific materialism. The first is consciousness:

It is just a plain fact about the world that it contains such conscious mental states and events, but it is hard to see how mere physical systems could have consciousness. How could such a thing occur? How, for example, could this grey and white gook inside my skull be conscious? (p. 15)

However, in Chapters 7 and 9 of the present book I show how wonderful scientific investigations resolve this mystery, at least at the first stage of understanding.

Searle states that it is

> something of a scandal that contemporary discussions in philosophy and psychology have so little of interest to tell us about consciousness. (p. 16)

Since that was written there has been a plethora of books and articles, some being quoted from in this chapter.

Now the ignorance about the microstructure and microfunctions of the brain is a much greater scandal. This knowledge is essential for any scientific understanding of the brain–mind problem (Chapters 4–9 of the present book) otherwise the brain indeed remains a meaningless grey and white gook!

Three other features besides consciousness make the mind–brain problem difficult—intentionality, subjectivity, and mental causation (p. 17).

So far I am in good general agreement with Searle; then unexpectedly he leaves what could be a dualistic belief system for the materialist belief that

> all of our mental life [is] caused by processes going on in our brain (p. 18)

and he repeats:

> Everything that matters for our mental life, all of our thoughts and feelings, are caused by processes inside the brain. (p. 19)

Notice the crudity of the brain reference. This is followed up by:

> Pains and other mental phenomena are just features of the brain and perhaps the rest of the central nervous system. (p. 19)

Searle is aware that this causal hypothesis will raise difficulties, so he relates it to other types of causation, for example liquidity or solidity, which seem to me to be unacceptable analogues.

My dualist concept is that mental events, such as intentions to move, are initially in the mental events relating to the cortical areas such as the supplementary motor area (Figure 9.5) in which neural activities are induced eventually (Figures 5.2 and 5.5) in bringing about the desired movements.

Searle's view at the end of his Chapter 1 is that mind and body interact, but they are not two different things, since mental phenomena just are features of the brain. At the end Searle retracts from this hard position with his concepts of naive physicalism and naive mentalism. He states that

> mental phenomena really exist and many of them function causally in determining physical events in the world.

It is good to have this assurance by Searle, but it is based on dubious neuroscience.

Chapters 3, 4, and 5 are not of vital interest to my brain–mind story, but reference should be made to Searle's Chapter 6 on freedom of the will. I agree with his statement:

> Our conception of human freedom is that it is essentially tied to consciousness. (p. 94)

However, the deterministic objection to free will is now eliminated by the rigorous treatment of Beck and myself (Beck and Eccles 1992 and Chapter 9 of the present book).

3.9 J. R. Searle, *The Rediscovery of the Mind*, 1992

It is surprising how close the themes in this book are to the theme of Searle's 1984 book, which have been commented on in the previous section. There is the same startling theme:

> How could this grey and white gook inside my skull be conscious?

Unfortunately an attempted answer to this question involves an understanding of the structure and function of the brain, particularly the neocortex (Chapters 4–10 of the present book), that goes far beyond the traditional neuroscience that Searle quotes (Kuffler and Nicholls 1976; Shephard 1983; Bloom and Lazerson 1988).

Searle's biological naturalism is expressed early in his book:

> What exactly is the character of the neurophysiological processes and how do the elements of the neuroanatomy—neurons, synapses, synaptic clefts, receptors, mitochondria, glial cells, transmitter fluids, etc.—produce mental phenomena both conscious and unconscious? (p. 1)

This is the question of an identity theorist who believes *that somehow material happenings in a complex system generate mental events* (Chapter 1.3.4 of the present book). But this statement is without a scientific basis in classical physics. Instead it can be interpreted as a request for magic!

Searle summarizes:

> Our world picture, though extremely complicated in detail, provides a rather simple account of the mode of existence of consciousness . . . Some types of living systems have evolved over long periods of time. Among these, some have evolved brains that are capable of causing and sustaining consciousness. (p. 93)

But, as just argued, this is a statement requiring magic.

In Chapter 2, 'The Recent History of Materialism', Searle states:

> I have been concerned not so much to defend or refute materialism as to examine its vicissitudes in the face of certain commonsense facts about the mind, such as the fact that most of us are, for most of our lives, conscious. What we *find* in the history of materialism is a recurring tension between the urge to give an account of reality that leaves out any reference to the special features of the mental, such as consciousness and subjectivity, and at the same time account for our 'intuitions' about the mind. It is, of course, impossible to do these two things. (p. 52)

Two quotations may give Searle's solution of the 'body–mind problem':

> If we had an adequate science of the brain, an account of the brain that would give causal explanations of consciousness in all its forms and varieties, and if we overcame our conceptual mistakes, no mind–body problem would remain. (p. 100)

> We do not know how the system of neurophysiology/consciousness works, and an adequate knowledge of how it works would remove the mystery (p. 102)

—at the expense of more magic!

Reference should be made to Chapter 7 (on the evolution of consciousness) of the present book for a scientific account of the evolution of consciousness that Searle may be able to accept. Because it brings in quantum physics it could help him in his removal of the magic. It includes subjectivity and can easily be extended to intentionality.

It is interesting that Searle states:

> If one had to describe the deepest motivation for materialism, one might say that it is simply a terror of consciousness. But should this be so? Why should materialists have a fear of consciousness? Why don't materialists cheerfully embrace consciousness as just another material property among others? (p. 55)

> The deepest reason for the fear of consciousness is that consciousness has the essentially terrifying feature of subjectivity. Materialists are reluctant to accept that feature because they believe that to accept the existence of subjective consciousness would be inconsistent with their conception of what the world must be like. Many think that, given the discoveries of the physical sciences, a conception of reality that denies the existence of subjectivity is the only one that it is possible to have. (p. 55)

I believe all of this amounts to a very large mistake, but this is because Searle accepts the identity theory, which I have just rejected!

Searle's Chapter 2 on the recent history of materialism is most valuable in giving and criticizing a whole variety of materialisms, including behaviourism, type identity theories, black box functionalism, strong artificial intelligence, and eliminative materialism.

Searle comes to an important conclusion:

> It would be difficult to exaggerate the disastrous effects that the failure to come to terms with the subjectivity of consciousness has had on the philosophical and psychological work of the past half century. In ways that are not at all obvious on the surface, much of the bankruptcy of most work in the philosophy of mind and a great deal of the sterility of academic psychology over the past fifty years, over the whole of my intellectual lifetime, have come from a persistent failure to recognize and come to terms with the fact that the ontology of the mental is an irreducibly first-person ontology. There are very deep reasons, many of them embedded in our unconscious history, why we find it difficult if not impossible to accept the idea that the real world, the world described by physics and chemistry and biology, contains an ineliminably subjective element. How could such a thing be? How can we possibly get a coherent world picture if the world contains these mysterious conscious entities? Yet we all know that we are for most of our lives conscious, and that other people around us are conscious. (p. 95)

I think that this represents an important contribution of Searle, and I fully support his repeated suggestion that what we urgently need is an adequate science of the brain.

It is important that Searle raises this stark picture with its sequel:

> . . . and unless we are blinded by bad philosophy or some forms of academic psychology, we really don't have any doubts that dogs, cats, monkeys and small children are conscious, and that their consciousness is just as subjective as our own. (p. 95)

Does Searle not realize that he too is blinded by bad philosophy and that he needs dualism to resolve his dilemma? Searle has rejected dualism because he links it to the two kinds of substances of Cartesianism. But I espouse the dualism of Popper that is illustrated by Worlds 1 and 2 of Figures 1.1 and 1.3.

One last comment is that I liked the coverpiece by van Gogh and would like to name it 'Two Selves Communing'!

3.10 R. W. Sperry, 'Forebrain Commissurotomy and Conscious Awareness, 1967

The great achievement of Sperry in studying patients with transection of the corpus callosum is not a controversial issue. Instead my concentration will be on the important controversy to do with the mind–brain problem. There is an immense literature. I select first his 1967 paper, which was chosen for the volume *Brain Circuits and Functions of the Mind*, composed of essays in honour of Sperry.

> In essence, consciousness was conceived to be a dynamic emergent of brain activity, neither identical with nor reducible to the neural events of which it is mainly composed. Further, consciousness was not conceived as an epiphenomenon, inner aspect, or other passive correlate of brain processing, but rather as an active integral part of the cerebral process itself, exerting potent causal effects in the interplay of cerebral operation. In a position of top command at the highest levels in the hierarchy of brain organization, the subjective properties were seen to exert control over the biophysical and chemical activities at subordinate levels. It was described initially as a brain model that puts 'conscious mind back into the brain of objective science in a position of top command . . . a brain model in which conscious, mental, psychic forces are recognized to be the crowning achievement . . . of evolution.' It must be emphasized at the outset that no direct empirical proof is available, any more than proof is available for the opposing behaviorist position. (p. 382)

Sperry claims that

> a conceptual explanatory model for psychoneural interaction is provided, stated in terms acceptable to neuroscience without violating the monistic principles of scientific explanation. The main focus is on the feature of causality. By all prior theories of consciousness, at all recognized by science, consciousness was interpreted to be acausal in brain function. (p. 384)

But this is not the case. Many years before (Eccles 1951, 1953), I was considering how intense dynamic activity in the immense neuronal networks of the cerebral cortex could be receptive to the mental influences of the will. Sperry states that

> our concept of the emergent subjective properties and their causality are still, of course, at a general abstract level. It remains by this and by any other theory thus far available to explain those critical organizational differences that distinguish brain processes with subjective properties from those without, and to define in exact operational and neural terms the essential functional role played by subjective awareness. (p. 385)

3.11 R. W. Sperry, *Science and Moral Priority*, 1982

Some years later Sperry wrote *Science and Moral Priority*. In the earlier chapters of this book he was particularly concerned with values and science, as in these quotations:

> I must take issue especially with the whole materialist-reductionist conception of human nature and mind that seems to emerge from the currently prevailing objective analytic approach in the brain-behavior sciences. When we are led to favor the implications of modern materialism in opposition to older idealistic values in these and related matters, I suspect that we have been taken, that science has sold society and itself a somewhat questionable bill of goods. (p. 28)
>
> Another serious threat to cherished images of human nature is the scientific rejection of free will. (p. 39)
>
> This does not mean, however, that there are cerebral operations that occur without antecedent cause. Man is not free from the higher forces in his own decision-making machinery. (p. 40)
>
> But to return to our central concern regarding the impact of creeping materialism in the brain-behavior sciences, and elsewhere, we can say in summary that it is possible to see today an objective explanatory model of brain function that neither contradicts nor degrades, but rather affirms, age-old humanist values, ideals, and meaning in human endeavor. The noble, free, or sublime qualities—or the opposite, for that is how meanings arise—that the humanist formerly thought he could see in man and his activities are present and sustained in our current scientific model, much as history and common experience have always shown. For those who like to perceive a take-home message, that of the present is quite simple, applying to scientist and humanist alike: Never underestimate the power of an ideal. (p. 44)

Later I find Sperry more controversial on the mind–brain action:

> A conceptual explanatory model is provided for the way mind can rule matter in the brain and exert causal influence in the guidance and control of behavior, on terms acceptable to neuroscience and without violating monistic principles of scientific explanation. (p. 66)
>
> No dualistic interaction in the classical sense is implicit. The causal power attributed to the subjective properties resides in the hierarchical organization of the nervous system and in the power exerted by any whole over its parts. Mind moves matter in the brain in much the same way that an organism moves its component organs and cells. (p. 66)

I fear that these crude analogical arguments are not acceptable scientifically. However, in Chapter 9 of the present book an explanation is given by Beck and myself in terms of quantum physics.

For our purpose the most interesting chapter is 'Mind–Brain Interaction. Mentalism, Yes; Dualism, No'. Sperry makes the important claim:

> In calling myself a mentalist, I hold subjective mental phenomena to be primary, causally potent realities as they are experienced subjectively, different from, more than, and not reducible to their physicochemical elements. At the same time, I define this position and the brain–mind theory on which it is based as monistic and see it as a major deterrent to dualism. (p. 79)

Of course I deny these statements. Dualism is misunderstood by Sperry, as I shall show in Chapter 9.

> In the mid-sixties, I did not venture to push them at the Vatican conference beyond mild reference to 'a view that holds that consciousness may have some operational and causal use'. To this Eccles responded by asking: 'Why do we have to be conscious at all? We can, in principle, explain all our input–output performance in terms of actvity of neuronal circuits; and consequently consciousness seems to be absolutely unnecessary!'. (p. 80)

Sperry apparently did not recognize that my comment was naughtily ironic! I was dedicated to the reality of consciousness with the title of my lecture 'Conscious Experience' and had first published on it in 1951, in *Nature*. The misunderstanding is amazingly evident when Sperry writes:

> I was delighted to see by his next International Brain Research Organization presentation that [Eccles] had clearly joined our ranks as an ardent antireductionist denouncing 'the materialistic, mechanistic, behavioristic, and cybernetic concepts of man.' Reversing his earlier stand on the uselessness of consciousness for a full account of brain function, Eccles has since lent his support to the new logic for the causal influence of mind over neural activity. On these points I believe we have remained in good general agreement. (p. 82)

Sperry claims that in 1952 he

> presented in terms of neuronal circuitry and concepts of neuroscience a theory that seemed to counter and refute, for the first time on its own grounds, the classic physicalist assumption of a purely physical determinacy of the central nervous system. Subjective mental phenomena had to be included. Mind–brain interaction was made a scientifically tenable and even plausible concept without reducing the qualitative richness of mental properties. The overall aim of the paper, as in the Popper and Eccles volume was to show that this recognition of the primacy of conscious mind as causal would alter profoundly the value implications of science which were being downgraded by the then strongly dominant philosophy of reductive mechanistic materialism. (pp. 81–82)

So I refer again to my 1951 *Nature* article!

In his proposal of mental monism Sperry is giving the cerebral cortex properties not recognized in classical physics, as Stapp has pointed out. He is in the same difficulty as all the identity theorists. It is important to recognize that in our philosophy of dualism Popper and I include much more than the self as existent over and above the brain. There is also the whole of Worlds 2 and 3.

Sperry finishes his book by a remarkable idealistic statement:

> Even for the immediate good of this, our own generation, it now becomes important that new, long-term, more godlike guidelines—of a kind that will insure long survival and further progress in the quality of life—be instituted very soon if humanity is to live again with a sense of hope, purpose or higher meaning. (p. 126)

I agree.

3.12 H. P. Stapp, 'Quantum Propensities and the Brain–Mind Connection', 1991

The quantum physicist Henry P. Stapp has for several years been thinking and publishing on the mind–brain problem and has been inspired by the concepts of William James, who wrote his monumental two-volume *Principles of Psychology* just over 100 years ago. Stapp has achieved a remarkable synthesis of the philosophy of William James with that of the theoretical physicist Werner Heisenberg into what Stapp calls the Heisenberg/James model of consciousness. These quotations present his views:

> A conscious thought is a real thing that has an *essential unity*. It is not merely an aggregation of simpler parts. It is fundamentally one whole thing.
>
> Nothing in classical physics can create something that is *essentially* more than an aggregation of its parts. But a quantum actual event does exactly that: it creates a single new actuality by grasping and combining together into a unified new ontological whole various diverse aspects of the prior situation. The availability of integrative actual events of this kind is one of the two fundamental reasons why one must turn to quantum theory to achieve a rationally coherent understanding of the mind–brain connection: the reductionist classical-physics conception of nature is logically unsuited to the task of accommodating essentially unified conscious thoughts.
>
> The second fundamental reason why we must turn to quantum theory is that classical physics has, as is well known, no rational place for consciousness: it is already logically complete. The physical world, as it is conceived in classical physics, consists of *nothing but* the various particles and fields, whose properties are, within that theory, spelled out completely. There is no logical place within this conceptual structure for another kind of entity

> such as conscious: if consciousness is put into the theory at all, then it must be put in simply 'by hand', rather than by virtue of the logical structure of the theory.
>
> The logical situation in quantum theory is quite different: there is an absolute logical need for something else, such as consciousness. (p. 1470)

This strong statement of Stapp has been of great value to me in giving reassurance in rejecting materialist reductionism and in building a dualist theory of the mind–brain problem as can be seen in Chapters 4–10 of the present book.

3.13 H. P. Stapp, 'A Quantum Theory of the Mind–Brain Interface', 1990

I refer again to Stapp's criticism of Edelman's book *The Remembered Present*, which I have extensively reviewed in Chapter 3.5. Stapp quotes Edelman:

> 'The functioning of these key [reciprocal] connections [between past value-category connections and current perceptual categorizations] provides the sufficient condition for the appearance of primary consciousness.' (p. 21)

Stapp goes on to state:

> The question arises as to how one is to interpret this claim that this special neural process is a *sufficient condition* for consciousness to occur. Does this claim mean that the occurrence of consciousness is *logically entailed* by the occurrence of this neural process?
>
> At the beginning of his book Edelman lists a set of constraints on his undertaking. The first of these is the condition that 'any adequate global theory of brain function must include a scientific theory of consciousness, but to be scientifically acceptable it must avoid the Cartesian dilemma. In other words, it must be uncompromisingly physical and be based on *res extensa*, and indeed be derivable from them.'
>
> This condition seems to demand that the emergence of consciousness be *derivable* from the properties of matter. Edelman accepts 'modern physics as an adequate description for our purposes of the nature of material properties'. Thus Edelman's demand appears to be that the emergence of consciousness must be actually derivable from physics, or at least from properties of systems describable in principle in terms of the concepts of physics. This strong interpretation is reinforced by the claim made in the final chapter that 'no special addition to physics is required for the emergence of consciousness'. (p. 21)
>
> If this indeed be the claim *then Edelman's account falls short.* (p. 22)

> The problem with Edelman's approach is that if one adheres to his demand that the 'view of brain function and consciousness should be based on materialist metaphysics', and hence rules out quantum physics, and perforce retreats to classical physics, then there is nothing in the *physics* that singles out these special processes as being in any way special. They are special only because they can be associated in a certain way with things outside classical physics, namely possible conscious experience. (p. 22)

3.14 H. P. Stapp, *Mind, Matter, and Quantum Mechanics*, 1993

I give some quotations to illustrate the Stapp–James interaction:

> Beyond this crucial issue of the efficacy of consciousness, James's principal claim, at the fundamental level, is the *wholeness, or unity, of each conscious thought.* (p. 11)

> Rather, consciousness is a comparatively simple aspect of the complex brain process. The availability to us of these glimpses, however flawed and fallible, into the complex workings of the brain provides scientists with insights that can be exploited. (pp. 13–14)

> William James, and other nineteenth-century psychologists, took consciousness to be the core subject matter of psychology, and introspection a necessary tool for investigating it. He recognized that 'introspection is difficult and fallible' and he apparently recognized that the problem of the connection of conscious process to brain process was irresolvable within the framework of the classical physics of his day. (p. 13)

It can be recognized that there is then attraction to Stapp about the identity theory in that it should give a unique insight 'into the complex workings of the brain'. The mistake of James was to believe that this insight would be delivered by introspection. Also the present cognitive psychology has been similarly disappointing. However, now there are exquisite scientific investigations on the brains of cooperating human subjects, as described in Chapters 4–9 of the present book. The future is illimitable. Stapp apparently makes a reference to this:

> Certain Heisenberg events that actualize large-scale patterns of neuronal activity in human brains will be identified as the physical correlates of human conscious events. (p. 20)

Stapp on p. 23 severely criticizes Dennett:

> Daniel Dennett's book *Consciousness Explained* approaches the problem of consciousness from the materialist point of view. He announces that 'It is

> one of the main burdens of this work to explain consciousness without ever giving in to the siren song of dualism. What, then is so wrong with dualism? Why is it in such disfavor? (p. 23)

See Chapters 4–9 of the present book.

References to Chapter 3

Changeux, J. P. (1985) *Neuronal Man: The Biology of Mind* (Pantheon, New York; first published by Fayard, 1983).

Crick, F., and Koch, C. (1990) Towards a neurophysiological theory of consciousness, *Seminars in the Neurosciences* **2**, 263–275.

Crick, F., and Koch, C. (1992) The problem of consciousness, *Scientific American* September 1992, pp. 111–117.

Dennett, D. C. (1991) *Consciousness Explained* (Allen Lane/Penguin, London).

Eccles, J. C. (1951) Hypotheses relating to the brain–mind problem, *Nature* **168**, 53–57.

Eccles, J. C. (1953) *The Neurophysiological Basis of Mind: The Principles of Neurophysiology* (Clarendon, Oxford).

Edelman, G. M. (1989) *The Remembered Present. A Biological Theory of Consciousness* (Basic Books, New York).

Hodgson, D. (1991) *The Mind Matters* (Clarendon, Oxford).

Lieberman, P. (1975) *On the Origins of Language* (Macmillan, New York).

Margenau, H. (1984) *The Miracle of Existence* (Ox Bow, Woodbridge CT).

Penrose, R. (1989) *The Emperor's New Mind: Concerning Computers, Minds, and the Laws of Physics* (Oxford University Press, Oxford).

Searle, J. R. (1984) *Minds, Brains and Science* (British Broadcasting Corporation, London).

Searle, J. R. (1992) *The Rediscovery of the Mind* (MIT Press, Cambridge MA).

Sperry, R. (1967) Forebrain commissurotomy and conscious awareness, in *Brain Circuits and Functions of the Mind*, edited by C. Trevarthen (Cambridge University Press, Cambridge), pp. 371–388.

Sperry, R. (1992) *Science and Moral Priority* (Columbia University Press, New York).

Stapp, H. P. (1990) A quantum theory of the mind–brain interface, in Stapp (1993).

Stapp, H. P. (1991) Quantum propensities and the brain–mind connection, *Foundations of Physics* **21**, no. 12, 1451. Reprinted in Stapp (1993).

Stapp, H. P. (1993) *Mind , Matter, and Quantum Mechanics* (Springer, Berlin, Heidelberg).

Stebbins, G. L. (1982) *Darwin to DNA, Molecules to Humanity* (Freeman, New York).

Tobias, P. V. (1983) Recent advances in the evolution of the hominids with special reference to brain and speech, in *Recent Advances in the Evolution of Primates*, edited by C. Chagas (Pontificiae Academiae Scientiarum Scripta Varia 50, Vatican City).

Wigner, E. P. (1967) *Symmetries and Reflections* (Indiana University Press, Bloomington IN).

4 New Light on the Mind–Brain Problem: How Mental Events Could Influence Neural Events

4.1 Introduction

It has long been recognized that, if non-material mental events, such as the intention to carry out an action, are to have an effective action on neural events in the brain, it has to be at the most subtle and plastic level of these events. Attention has to be focused on the biological units of the brain, the neurons or nerve cells, and on the manner of their communication at specialized sites of close contact, the synapses. An introduction to conventional synaptic theory leads on to an account of the manner of operation of the ultimate synaptic units. These units are the synaptic boutons that, when excited by an all-or-nothing nerve impulse, deliver the total contents of a single synaptic vesicle, not regularly, but probabilistically. *This quantal emission of synaptic transmitter molecules* (about 5 000 to 10 000) is the ultimate functional unit of the transmission process from one neuron to another. This refined physiological analysis leads on to an account of the ultrastructure of the synapse, which gives clues as to the manner of its unitary probabilistic operation. The essential feature is that the effective structure of each synapse is a *paracrystalline presynaptic vesicular grid*, which acts probabilistically in quantal release.

When considering mental events in relation to a possible influence on these subtle neural events, it is essential to avoid mental events with levels of complexity that result in a confusion of neural happenings which are beyond analysis. There will be in Chapters 5, 6, and 10 an account of several recent studies in which mental events result in such simple neural events that correlation with probabilistic quantal operation of synapses is possible. Conventional neurophysiology enables one to transcend unitary synaptic actions by operations of the known neural pathways, so that a mental intention to move can become fulfilled by the desired movement.

In the final stage of this enquiry it has to be considered how a non-material mental event, such as an intention to move, can influence the subtle probabilistic operations of synaptic boutons. On the biological side, attention will be focused on the paracrystalline presynaptic vesicular grids as the targets for non-material mental events. On the physical side, attention will be focused on the probabilistic fields of quantum mechanics, which carry neither mass nor energy, but which nevertheless can exert effective action at microsites. The new light on the mind–brain problem comes from the hypothesis that the non-material mental events, the World 2 of Popper, relate to the neural events of the brain (the World 1 of matter and energy) by actions in conformity with the physics of quantum theory (Chapter 9). *This hypothesis opens up an immense field of scientific investigations both in quantum physics and in neuroscience.*

4.2 Quantum Insights into the Mind–Brain Problem

It is generally recognized that progress in the understanding of the mind–brain problem and the nature of consciousness depends on great advances in the understanding of the brain. For example, in Chapter 3 reference can be made to Searle, Hodgson, Penrose, Dennett, and Crick and Koch. Yet this progress lags because of the belief that the brain is a supercomplex electronic device. It is studied in all the artifical intelligence investigations with the neural network computations and robotic modelling of Minsky, Moravec, Edelman, and Changeux. I think this is a great error arising from the failure to study the necessary microlevels of both the structure and function of the neocortex.

As can be realized, in Chapters 4–10 of this book, it is not necessary for *neurophilosophers* to study the whole of neuroscience, which would be a most daunting task. But there is no text giving the basic information, which would have been most valuable for Penrose, Hodgson, and Searle for example. So I give at the start of this chapter important knowledge on the synapse, which is the fundamental unit of cerebral activity. The more difficult sections on the quantum insights are set in small type and can be used for reference.

The new quantum approach to dualism is described. For the first time, I recognized the fundamental significance of Akert's beautiful paracrystalline structure of the presynaptic vesicular grid (Figure 4.5) of the synaptic boutons with its low probability for quantal emission of transmitter in exocytosis.

At the end of Chapter 2 I described the impact of Margenau's 1984 book, which provided the inspiration I had needed in my long quest for a dualist solution of the mind–brain problem.

This chapter gives my first attempt to present a detailed study of synaptic microneuroscience, which could open the opportunity for the operation of the probability fields that were essentially involved in Margenau's application of quantum physics. There is an account of the units of operation of the cerebral cortex, the synapses with the presynaptic vesicular grids, and the control of vesicular transmissions by the exocytoses. *The detailed account here is justified because it concerned essential neuroscience that had not before been appreciated* but that turned out to be fundamental for further progress in utilizing quantum physics for the mind–brain problem, as can be realized in later chapters.

What was missing was, firstly, how a mental event could select for an exocytosis from the presynaptic vesicular grid of a cortical synapse; and, secondly, a detailed description of how the exocytosis could participate in an amplifying system of parallel actions whereby the milli-EPSPs generated by one exocytosis could become summed so as to be effective in neural action through triggering impulse discharges from the pyramidal cells. This requires more information of the neuronal structures of the neocortex. The existing knowledge of the modular arrangement of cortical neurons, as diagrammed by Szentágothai (Figures 6.3b and 6.5), lacked the needed great amplification in a built-in design that is described in Chapter 6, the dendron story.

4.3 The Integrative Action of Ia Impulses on a Motoneuron

A simple introduction to the synaptic concept of brain action is provided by an account of the mode of action of impulses in Ia nerve fibres on a motoneuron as shown diagrammatically in Figure 4.1a. These large nerve fibres come from the annulospiral endings in a muscle and directly excite the motoneurons of that muscle as indicated by the intracellular recordings in Figure 4.1b–j. This is the neuronal system responsible for the simple knee jerk. As the stimulus applied to the bundle of Ia fibres was progressively increased to excite more and more of the nerve fibres (see the sharp upper traces of nerve spikes) converging on that motoneuron, there was a corresponding increase in the brief depolarizations recorded intracellularly (note the ms time scale). The intracellular microelectrode shown diagrammatically in Figure 4.1a recorded a resting potential across the membrane of the motoneuron of about −70 mV (internal negativity), and synaptic stimulations caused a brief diminution of this resting potential, the excitatory postsynaptic potential, which increased up to 7 mV in Figure 4.1g–j when all the Ia fibres were stimulated. In Figure 4.1g–j there was an apparent discrepancy, the nerve fibre volley increasing with no corresponding increase

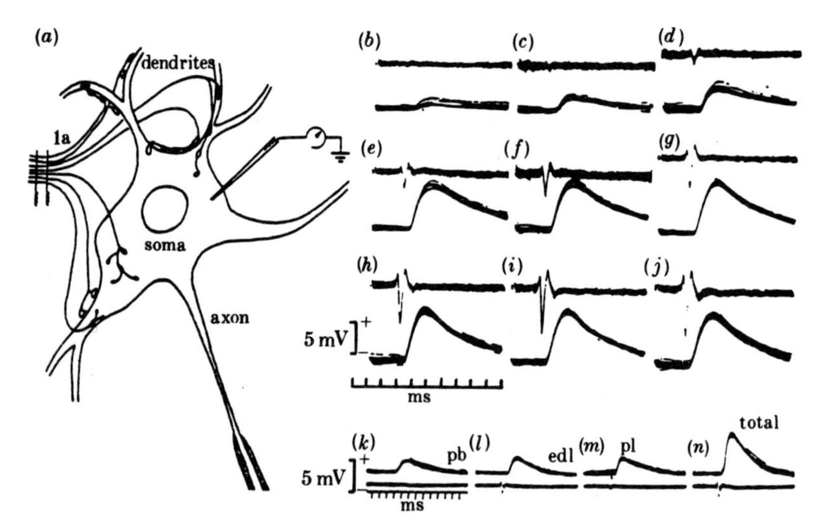

Figure 4.1 Monosynaptic excitation of motoneurons by the group Ia afferent pathway. (**a**) A drawing of a motoneuron showing the central dendritic regions, the soma, the initial segment of the axonal origin, and the beginning of the axonal medullation. On the dendrites and soma are shown the excitatory synaptic endings of seven group Ia afferent fibres that have an applied stimulating electrode (actually in the peripheral muscle nerve). The intracellular microelectrode recording is shown diagrammatically. (**b–j**) The *upper traces* give the size of the afferent volley as it enters the spinal cord, and the *lower traces* the simultaneously recorded EPSPs. All records are formed by the superimposition of about 25 faint traces. (**k–m**) The EPSPs recorded in another motoneuron (peroneus longus) in response to maximum group Ia volleys in the nerves to three muscles—peroneus brevis, extensor digitorum longus, and peroneus longus. (**n**) All three muscles combined. (Eccles et al. 1957).

in the EPSP, but this was due to the excitation of another class of nerve fibres (Ib) that do not contribute to the monosynaptic EPSP. Figure 4.1k–n shows that the Ia EPSPs sum linearly when small: (n) = (k) + (l) + (m).

When the bundle of Ia fibres shown in Figure 4.1a was cut down to one, the EPSPs were very small but could be amplified by successive addition in a computer. In that way it was shown that a single Ia fibre was distributed very widely to motoneurons of its muscle of origin. For example, in Figure 4.2a the same group Ia fibre produced EPSPs in six different motoneurons (Mendell and Henneman 1971). By a double horseradish peroxidase (HRP) technique it has been possible to identify all the synapses made by a single Ia fibre on a motoneuron (Burke et al. 1979; Brown 1981). For example, the diagram of Figure 4.2b represents the locations of synaptic endings by a single Ia afferent from medial gastrocnemius muscle on a medial gastrocnemius motoneuron. The five synaptic endings are on three different dendrites, two being rather close to the soma, the others more distant. Wide ranges of distribution are found, but there is a tendency for clustering. This will account

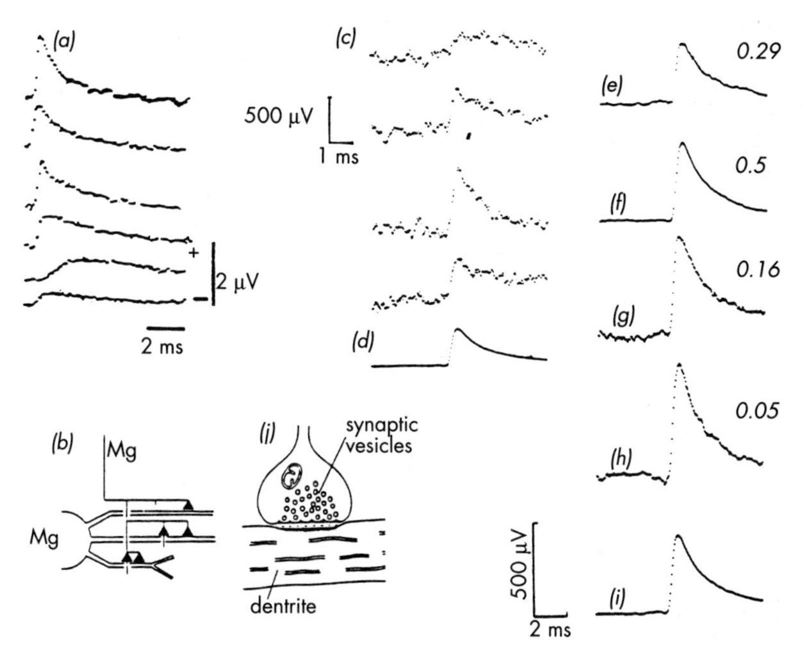

Figure 4.2 (**a**) Averaged recordings of EPSPs produced by impulses in the same Ia fibre terminating on six different motoneurons (Mendell and Henneman 1971). (**b**) A summary diagram of the location of Ia synapses from a single medial gastrocnemius Ia fibre on to a medial gastrocnemius motoneuron at five sites on three different dendrites as indicated (Brown 1981). (**c**) Four individual EPSPs selected from a population of 800 responses. (**d**) The average of all the 800 responses. (**e**) The component 1 of the EPSP derived from fluctuation analysis. (**f–h**) The components 2, 3, and 4 of this same fluctuation analysis. The probabilities of the occurrence of these components are indicated to the right of each. (**i**) The reconstructed EPSP obtained by adding the weighted sum of (e), (f), (g), and (h): 0.29(e)+0.5(f)+0.16(g)+0.05(h). (Jack et al. 1981a). (**j**) A drawing of a synapse on a dendrite to show the bouton with vesicles and the synaptic cleft.

for the range in time course of EPSPs produced by a single Ia fibre in Figure 4.2a. On some motoneurons there was clustering of its boutons close to the soma, on others more dispersal, and on the second lowermost trace there would be a synaptic clustering far out on dendrites.

4.4 The Quantal Emission of a Bouton

A still more refined level of enquiry concerns the EPSP produced by a single bouton, as indicated in Figure 4.2j, which is a component of the several bouton endings of a single Ia fibre (Figure 4.2b). By a technique of fluctuation analysis (Redman 1980; Hirst et al. 1981; Jack et al. 1981a, 1981b) it has been possible to distinguish between the EPSPs generated by each bouton on a motoneuron when activated by a single Ia impulse. For example, Figure 4.2c illustrates the wide range of fluctuating EPSPs produced by a single Ia impulse, and in Figure 4.2d there is summation of 800 such responses to give a typical unitary EPSP, such as those of Figure 4.2a. By the fluctuation analysis this EPSP is shown to be composed of elements, each generated by a single bouton. It is very rare that an activated bouton liberates more than one vesicle, which is the quantal package of the synaptic transmitter (cf. Figure 4.2j). In Figure 4.2e–h there are shown 4 EPSPs derived by the fluctuation analysis of Figure 4.2d, each arising from a single bouton. The probability of quantal emission from a bouton of a single synaptic vesicle ranges from 0.5 to 0.05, Figure 4.2e–h showing the time courses of the EPSPs generated by the synaptic emission from each bouton. In sequence for Figure 4.2 the sizes of the quantal EPSPs are 302, 406, 505, and 607 μV and when they are summed with allowance for probability there is an accurate reconstruction of the EPSP produced by a single fibre, Figure 4.2i, which is identical with Figure 4.2d when allowance is made for the different voltage scales. It can be assumed that the four derived EPSPs of Figure 4.2e–h are each produced by a single bouton at various distances from the soma, Figure 4.2h closest, Figure 4.2e most remote, as in Figure 4.2b.

From this remarkable analysis (Jack et al. 1981a) there are derived two conclusions on the presynaptic functioning of a single Ia fibre on a motoneuron:

1 There is a wide gradation of intermittency, for example, 0.5–0.05 in Figure 4.2e–h. Some boutons may even approach a probability of 1, but above 1 is not observed.
2 Usually an Ia fibre gives 3–5 boutons to a motoneuron (cf. Figure 4.2b), but the observed range is 1–10.

Korn and associates (Korn et al. 1982; Korn and Faber 1985) have studied a very different synapse, the inhibitory synapses on the Mauthner cell in the fish spinal cord (Figure 4.3a). It was possible to carry out a fluctuation analysis of the inhibitory postsynaptic potentials (IPSPs) produced by a single presynaptic inhibitory fibre (Figure 4.3b). They employed a different technique from that used by Jack, Redman, and associates (Figure 4.2), a binomial analysis, which showed a composition of 6 quanta (Figure 4.3c) in the response amplitude, and this number is in agreement with the histologically determined number of boutons (Figure 4.3c, d).

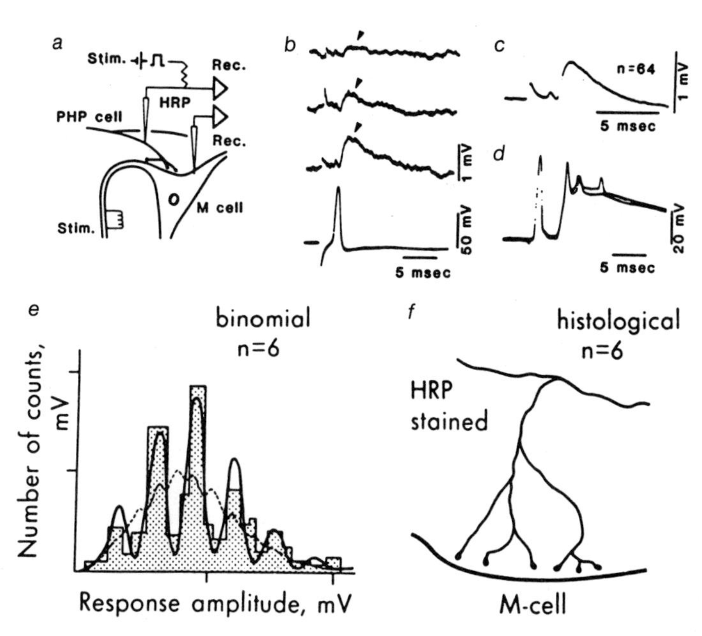

Figure 4.3 Evidence for quantal fluctuations of unitary IPSPs. (**a**) The experimental arrangement used for simultaneous intracellular recordings (Rec) from the M cell and a presynaptic inhibitory interneuron (PHP cell), both neurons being identified by their characteristic responses to antidromic stimulation (Stim.) of the M axon in the spinal cord. (**b**) Properties of depolarizing IPSPs recorded in a Cl-injected M cell throughout the same experiment. The variable amplitude of unitary IPSPs (*arrows* in *upper three traces*) following single presynaptic impulses directly evoked at a frequency of 1 per second. Only one presynaptic spike is shown (*lower trace*). (**c**) A computer-averaged unitary IPSP ($N = 64$). (**d**) The maximum-amplitude IPSP following antidromic activation of the recurrent collateral network was large enough to fire the M cell (Korn et al. 1982). (**e**) Correlation of mathematical and histological results provided in the same experiment. Following successive stimulations of a physiologically identified interneuron (rate: 1 per second), the resultant amplitude histogram of fluctuating unitary IPSPs (*shaded*) was analysed with a computer program that, taking noise into consideration, gave the best possible fits based upon the theoretical Poisson (*dashed line*) and binomial (*continuous curve*) equations. Obviously, the latter provided a better approximation. The six peaks correspond to the binomial term n, which defines the number of quanta. (**f**) A drawing of the camera lucida reconstruction of terminal arborization of the investigated HRP-filled presynaptic cell, indicating that the number of boutons established on the M-cell (histological n) was equivalent to the binomial one. This type of result led to the conclusion that each terminal knob is an all-or-none releasing unit. (Korn and Faber 1985)

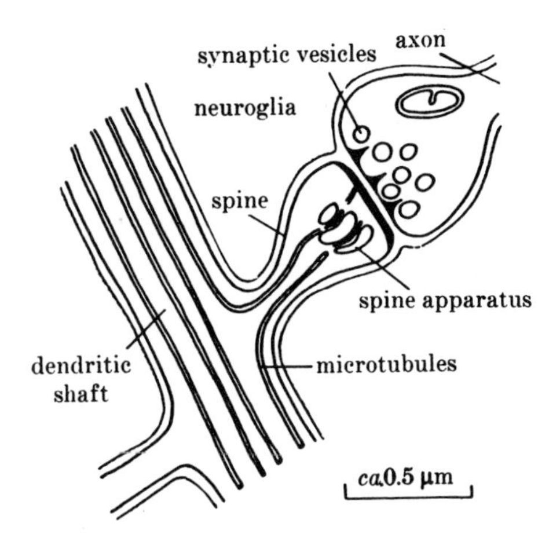

Figure 4.4 A drawing of a synapse on a dendritic spine. The bouton contains synaptic vesicles and dense projections of the presynaptic membrane (Gray 1982).

4.5 The Structure of a Chemically Transmitting Synapse in the Central Nervous System of Vertebrates

A synapse is usually formed by an expansion of a fine nerve fibre to form a *bouton* that makes a close contact with the surface of a dendrite or soma of a neuron. Figure 4.4 shows the commonest type of synapse, where the bouton makes contact across a narrow space, the synaptic cleft ($\sim$ 200 Å), with the expanded end of a dendritic spine (Gray 1982). Noteworthy are the synaptic vesicles and the dense triangular projections about 1000 Å apart from the presynaptic membrane into the bouton (10 Å = 1 nm.).

It is important to state at the outset that, in accordance with Hubbard (1970), the synaptic vesicles observed in all boutons in the central nervous system of vertebrates are the morphological correlates of the quantal emission of transmitter that occurs in all chemically transmitting synapses made by boutons on nerve cells. The synaptic vesicles are recognized as quantal packages of the preformed transmitter molecules (about 5 000 to 10 000) that are ready for release as a quantal package into the synaptic cleft in a unitary operation (Figure 4.4).

Subsequent study by the freeze-etching technique (Akert et al. 1975) confirmed and clarified the original findings of Figure 4.4, so that an idealized bouton could be drawn in perspective (Figure 4.5), with to the left the triangular array of the dense projections (az) ordering the hexagonal

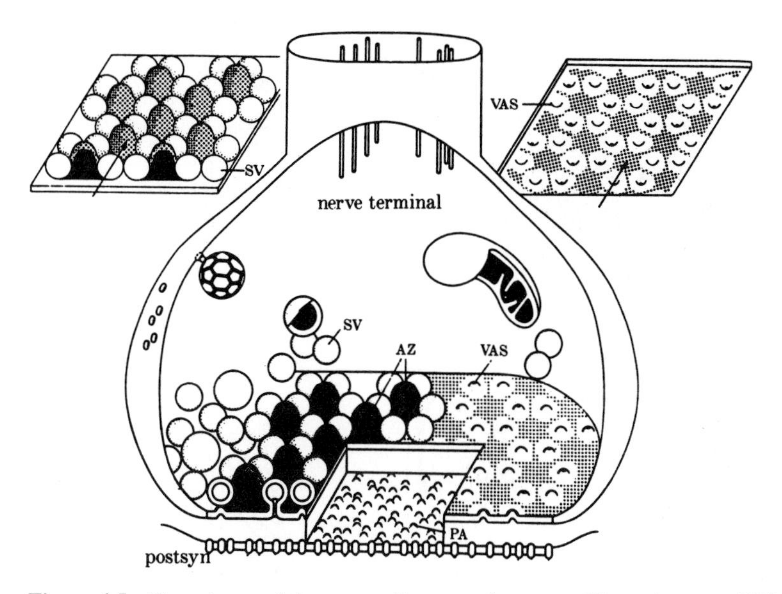

Figure 4.5 The schema of the mammalian central synapse. The active zone (AZ) is formed by presynaptic dense projections. The postsynaptic aggregation of intramembraneous particles is restricted to the area facing the active zone. (SV = synaptic vesicles; PA = particle aggregations on postsynaptic membrane (postsyn). Note the synaptic vesicles (SV) in hexagonal array, as is well seen in the *upper left inset* and the vesicle attachment sites (VAS) in the *right inset*. (See the text for further description.) (Akert et al. 1975)

array of synaptic vesicles (sv) confronting the synaptic cleft in a paracrystalline structure. Other vesicles lie further back, being in reserve. On the right side, stripping of the grid shows the vesicle attachment sites (vas) to the presynaptic membrane, while in the centre window are seen across the synaptic cleft the particle aggregations (pa) that probably relate to the transmitter receptors on the postsynaptic membrane. On each side of the nerve terminal there are shown in rectangular perspective the presynaptic vesicular grid and the presynaptic membrane after it has been stripped off.

Figures 4.4 and 4.5 show synapses from the mammalian brain. Essentially similar synapses are observed in a very different neuronal system, the inhibitory synapses on the Mauthner cell in the fish spinal cord (Triller and Korn 1982).

Figure 4.5 illustrates an important feature, namely that as a rule a synaptic bouton has only one presynaptic vesicular grid which may cover only a fraction of the total area of synaptic contact.

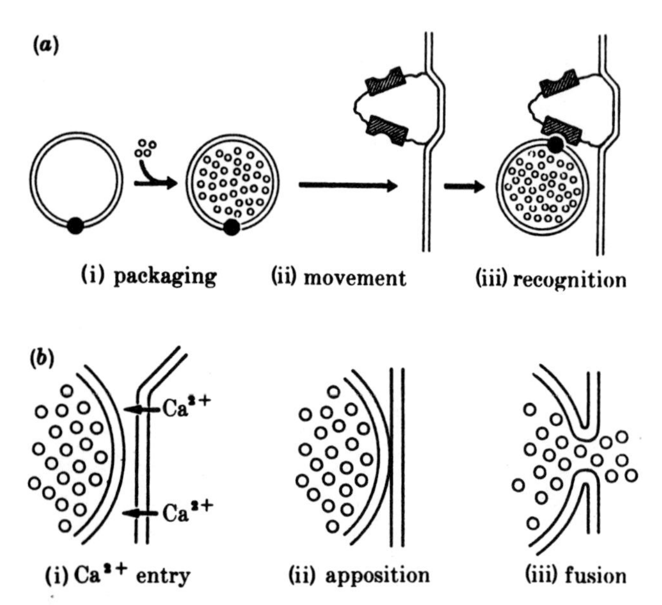

Figure 4.6 Stages of synaptic vesicle development, movement, and exocytosis. (**a**) The three steps involved in filling a vesicle with transmitter and bringing it to attachment to a presynaptic dense projection of triangular shape. (**b**) Stages of exocytosis with release of transmitter into the synaptic cleft, depicting the essential role of Ca^{2+} input from the synaptic cleft (Kelly et al. 1979).

There are still great problems (Figure 4.6) involved in the origin of synaptic vesicles, their charging by the specific transmitter molecules, the movement through the bouton for incorporation in the presynaptic vesicular grid, which we may call the firing zone (Figure 4.5), and the paracrystalline properties of this grid whereby of the large number of its constituent synaptic vesicles no more than one is emitted by a nerve impulse (Figures 4.5 and 4.6b) and that only with a probability usually far below one.

After this introduction to the fine structure of a bouton we have to consider how it comes about that, from this relatively large population of synaptic vesicles on the firing line of the presynaptic vesicular grid (Figure 4.5), rarely, if ever, is more than one discharged by a presynaptic impulse, and usually the probability is 0.5 or less (Jack et al. 1981a; Korn et al. 1982; Korn and Faber 1985; Redman and Walmsley 1983b). I think it has to be proposed that the presynaptic vesicular grid (cf. Figures 4.5 and 4.6) is a subtle dynamic structure designed to limit the emission of synaptic vesicles by a presynaptic impulse. By its unitary operation it sets the probability of vesicular emission from its whole structure and thus from the bouton.

It has to be remembered that there are probably no more than 2000 synaptic vesicles in a bouton, which is adequate for only a few minutes of normal operation.

According to the deconvolution analysis of Redman and associates (Jack et al. 1981a; Redman and Walmsley 1983b) the probability of quantal release may be almost one for some boutons and down to zero for others, with all ranges in between. According to the analysis of Korn and associates (Korn et al. 1982; Korn and Faber 1985) the mean probability for the different synapses ranged from 0.17 to 0.62. Various factors modify the probability, the most important being the increase due to the Ca^{2+} ion (Figure 4.6b), which acts after combination with the protein calmodulin, four Ca^{2+} ions to one molecule (de Lorenzo 1981). Other factors are the enhancing effect by a prior stimulus (Hirst et al 1981; Jack et al. 1981b), probably owing to the residual Ca^{2+} in the bouton, and by 4-aminopyridine (Jack et al. 1981b) and the depressant effect of high-frequency stimulation, halving at about 33 Hz (Korn and Faber 1985).

Despite the general agreement that the probability of release of a vesicle (quantum of transmitter) from a bouton for a single presynaptic impulse is less than unity, there are minor conflicts between the findings of Redman and associates and of Korn and associates.

These derive from the assumptions used in the deconvolution techniques from the recorded EPSPs. Redman's group (Jack et al. 1981a; Redman and Walmsley 1983b) assumes that the EPSPs produced by quantal emission from the boutons of an Ia fibre on a motoneuron are of a standard size, the deconvolution on this basis giving the probability of emission for each bouton. This analysis yields a bouton number usually much less than that actually determined by horseradish peroxidase analysis, so the assumption has to be made that some boutons have a zero probability of emission. On the other hand, Korn's group (Korn et al. 1982; Korn and Faber 1985) used binomial theory in their analysis of the synaptic responses of a Mauthner cell to a single inhibitory impulse and found a very good quantitative relationship between the analytical bouton number and the histological number (cf. Figure 4.3c, d); hence they assume the correctness of this deconvolution technique, according to which all the bouton endings of a single axon on a Mauthner cell have the same probability of emission of a single vesicle, the mean probability value being 0.38 for the 18 axons studied.

In conclusion I would suggest that neither deconvolution technique is fully acceptable.

The analysis by Redman and associates leads to a highly improbable solution of bouton number, with many boutons at zero quantal probability. The solution also requires that a quantal emission activates all the transmitter sites on the postsynaptic membrane. The solution of Korn and associates is superficially the most attractive, because the binomial analysis gives results in good accord with the number counted in horseradish peroxidase preparations (Figure 4.3c, d). However, the assumption of identical probabilities for all the boutons of an axon seems unlikely.

It is fortunate that both deconvolution techniques deliver essentially the same information, namely that *in response to a single presynaptic impulse a single synaptic bouton is a quantal emitter of vesicles with a probability usually well below one and never above one.*

This refined story of the quantal operation of synaptic boutons indicates that there should be a reexamination of the responses of neurons at the highest level of the central nervous system for which there is evidence of responses evoked by a mental intention or mental attention (Figures 5.2b, 5.4a, and 5.4b).

4.6 Introduction to the Mind–Brain Problem

There are many materialist theories of the mind, as summarized in the four entries of Figure 1.2. Radical materialism eliminates itself. The three other materialist theories recognize the existence of mind, or mental events, but give it no independent status. According to the above three materialist theories of the mind, mental states are an attribute of matter or the physical world, either of all matter as in panpsychism, or of matter in the special state in which it exists in the highly organized nervous systems of animals and man. One variety of this, epiphenomenalism, need not be further considered, having been replaced in recent decades by the identity theory that was first fully developed by Feigl (1967). Popper, in Popper and Eccles (1977), states that

> all four assert that the physical world (World 1) is self-contained or *closed . . . This physicalist principle of the closedness of the physical World 1 . . . is of decisive importance . . . as the characteristic principle of physicalism or materialism.*

Popper then goes on to give a critical account of all materialistic theories of the mind. Figure 1.1 gives diagrammatically Popper's three-world system.

It has been difficult to discover statements by philosophers that relate to the precise neural events which are assumed to be identical with mental events. The clearest expression was given by Feigl (1967). On p. 79 he states:

> The identity thesis which I wish to clarify and to defend asserts that the states of direct experience which conscious human beings 'live through', and those which we confidently ascribe to some of the higher animals, are identical with certain (presumably configurational) aspects of the neural processes in those organisms . . . processes in the central nervous system, perhaps especially in the cerebral cortex . . . The neurophysiological concepts refer to complicated highly ramified patterns of neuron discharges.

We can raise the question whether there could be experimental testing of predictions from the dualist-interactionist hypothesis on the one hand and the identity hypothesis on the other.

4.7 The Mind–Brain Problem

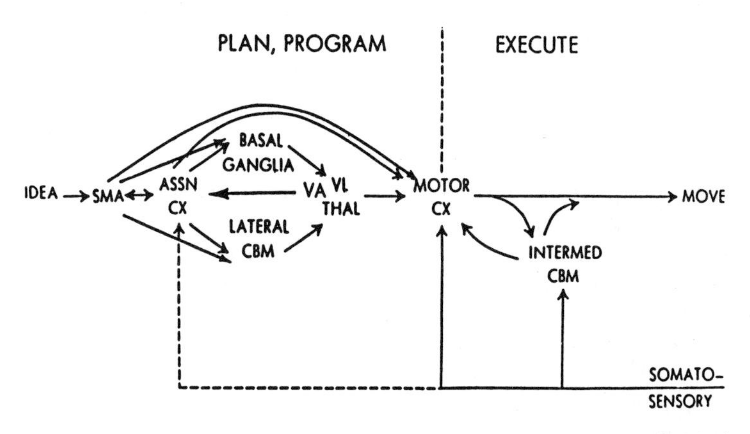

Figure 4.7 A diagram showing the pathways concerned in the execution and control of voluntary movement. ASSN CX = association cortex; LATERAL CBM = cerebellar hemisphere; INTERMED CBM = pars intermedia of cerebellum. The *arrows* represent neuronal pathways composed of hundreds of thousands of nerve fibres. SMA = supplementary motor area. From the motor cortex (MOTOR CX) the *arrow* to move is the pyramidal tract.

The identity theorist is committed to the doctrine that mental events per se cannot contribute to the generation of neural events (Figures 5.2b and 5.4a, b), which is the doctrine of *the closedness of World 1*. The experiences of an intention to move must therefore be attributable to the activity of an assemblage of neurons in the SMA (cf. Figures 4.7, 5.1, and 5.5). One may ask, what causes these neurons to fire impulses and so bring about a voluntary movement? To this question the identity theorist has no answer except to say that there must be other as yet unidentified neuronal centres that fire before the SMA, which is merely an evasion of the problem. This is the problem considered in Chapters 9 and 10, with illustration in Figures 9.5 and 10.2.

By contrast, the experimental findings are in accord with predictions from dualist-interactionism. As indicated in Figure 4.7 by the arrows, each mental event of intention (Figure 5.5) initiates (shown by arrows) the firing of a set of neurons of the SMA that, through the various known pathways indicated diagrammatically in Figure 4.7, cause the 'correct' motor cortical neurons to discharge impulses down the pyramidal tract to cause the desired voluntary movement (Eccles 1982a, 1982b).

The specific cortical activities produced by attention (Figure 5.4) in the absence of all neural inputs or overt activity of the cerebral cortex are comparable to the effects of internal programming of intention (Figure 5.2b) and likewise present an apparently inexplicable explanatory problem for the identity theorist. The dualist-interactionist is able to account for the action of the mental event of attention on the neurons of the prefrontal lobe as indicated in Figure 5.4. It is assumed, but not proven, that this is the site of the interaction, and that the neurons of the touch area of the cerebral cortex were secondarily activated (Roland 1981). A similar diagram could be made for the action of internal counting on the medial prefrontal cortex (Figure 5.4b).

However, the materialist critics argue that insuperable difficulties are encountered by the hypothesis that immaterial mental events can act in any way on material structures such as neurons. Such a presumed action is alleged to be incompatible with the conservation laws of physics, in particular of the first law of thermodynamics. This objection would certainly be sustained by 19th-century physicists, and by neuroscientists and philosophers who are still idealogically in the physics of the 19th century, not recognizing the revolution wrought by quantum physicists in the 20th century. Unfortunately it is rare for a quantum physicist to dare an intrusion into the brain–mind problem. But in a recent book the distinguished quantum physicist Margenau (1984) makes a fundamental contribution. It is a remarkable transformation from 19th-century physics to be told that

> some fields, such as the probability field of quantum mechanics, carry neither energy nor matter. (p. 22)

In summary, the mind may be regarded as a field, in the accepted physical sense of the term. But it is a non-material field; its closest analogue is perhaps a probability field.

4.8 Summary

Attention is concentrated on the mode of operation and the structure of the unitary transmitting elements of the brain, the synaptic boutons. Intracellular records of the responses of a neuron to an impulse in a single presynaptic fibre have been analysed to reveal the response evoked by a single bouton. This has been shown to act in a quantal manner, corresponding to the emission of a single vesicle of transmitter. However, the response is probabilistic, the emission occurring with a probability of usually 0.5 or less and never above 1.0. The microstructure of the bouton reveals that those synaptic vesicles awaiting exocytosis are embedded in a paracrystalline structure, the presynaptic vesicular grid, which in an unknown manner acts globally to ensure that the probability of emission is below 1 for the ensemble of 40 or more vesicles embedded in it. There is only one such grid for a bouton in the central nervous system.

The hypothesis of mind–brain interaction is that mental events act by a quantal probability field to alter the probability of emission of vesicles from presynaptic vesicular grids. There must be an immense operation in parallel on the thousands of presynaptic vesicular grids confronting a neuron with, in addition, many neurons similarly activated. Then by conventional neuronal circuitry mental events of intention achieve the desired brain response, leading to the desired motor movements. There is a similar explanation of the action of concentrated mental attention activating special areas of the cerebral cortex. Thus the mind–brain interaction postulated in dualism has been shown to be in accord with quantum physics. It does not violate natural laws as has been maintained by its critics. These concepts are fully supported in Chapter 9.

5 Do Mental Events Cause Neural Events Analogously to the Probability Fields of Quantum Mechanics?

5.1 Introduction

If non-material mental events, such as the intention to carry out an action, are to have an effective action on neural events in the brain, it has to be at the most subtle and plastic level of these events, as outlined in Chapter 4.

In the first stage of our enquiry an introduction to conventional synaptic theory leads on to an account of the manner of operation of the ultimate synaptic units. These units are the synaptic boutons that, when excited by an all-or-nothing nerve impulse, deliver the total contents of a single synaptic vesicle, not regularly, but probabilistically. This quantal emission of the synaptic transmitter molecules (about 5 000–10 000) is the elementary unit of the transmission process from one neuron to another.

In the second stage this refined physiological analysis leads on to an account of the ultrastructure of the synapse, which gives clues as to the manner of its unitary probabilistic operation. The essential feature is that the effective structure of each bouton is a paracrystalline presynaptic vesicular grid with about 50 vesicles, which acts probabilistically in vesicular (quantal) release.

In the third stage it is considered how a non-material mental event, such as an intention to move, could influence the subtle probabilistic operations of synaptic boutons. On the biological side, attention is focused on the paracrystalline presynaptic vesicular grids as the targets for non-material mental events. On the physical side, attention is focused on the probabilistic fields of quantum mechanics, which carry neither mass nor energy, but which nevertheless can exert effective action at microsites. The new light on the mind–brain problem came from the hypothesis that the non-material mental events, the 'World 2' of Popper, relate to the neural events of the brain, the 'World 1' of matter and energy, by actions in conformity with quantum theory. This hypothesis that mental events act on probabilistic synaptic events in a manner analogous to the probability fields of quantum mechanics seems to open up an immense field of scientific investigation both in quantum physics and in neuroscience.

All attempts to formulate a dualist hypothesis of brain–mind interaction are met with the strong criticism that such a hypothesis violates the conservation laws of physics. On this basis it is maintained that the world of matter–energy (the World 1 of Popper) is completely closed to the action of any non-material agency such as the self (the World 2 of Popper). These critics as a rule do not deny their mental experiences. They are dualists of a sort, but they deny the effectiveness of mental events, such as an intention to move, in causing or modifying neural events in the motor centres of the brain (see Chapters 1–3). There are many versions of such parallelist or identity or physicalist theories in which it is proposed that the mental events 'somehow' are identical with a special class of neural events, as was first proposed by Feigl (1967). *Such theories lack precision* in their formulations but have been generally accepted because they do not violate the *closedness of World 1.*

Similarly the dualist-interactionist hypothesis of Popper and myself (1977) (see Chapter 1.4) lacked a precise formulation of the site and manner of the postulated mental–neural interaction. Its appeal lay in its explanatory power of our experiences, particularly in respect of voluntary movement, where we seem indubitably to bring about actions at will.

5.2 The Possible Role of a Non-material Mental Event Acting at Brain Microsites Analogously to the Probability Fields of Quantum Mechanics

In a recent book the quantum physicist Margenau (1984) has suggested that a non-material mental event such as an intention to move could influence neural events at microsites without violating the conservation laws of physics. He states:

> In very complicated physical systems such as the brain, the neurons and sense organs, whose constituents are small enough to be governed by probabilistic quantum laws, the physical organ is always poised for a multitude of possible changes, each with a definite probability; if one change takes place that requires energy, or more or less energy than another, the intricate organism furnishes it automatically. Hence, even if the mind has anything to do with the change, that is, if there is a mind–body interaction, the mind would not be called upon to furnish energy. (p. 96)

In summary, Margenau states:

> The mind may be regarded as a field in the accepted physical sense of the term. But it is a nonmaterial field; its closest analogue is perhaps a

> probability field. It cannot be compared with the simpler nonmaterial fields that require the presence of matter (hydrodynamic flow or acoustic) . . . Nor does it necessarily have a definite position in space. And so far as present evidence goes it is not an energy field in any physical sense, nor is it required to contain energy in order to account for all known phenomena in which mind interacts with brain. (p. 97)

Hitherto such postulated microsites have had no specific identification, but now the evidence presented in Chapters 4.2 and 4.3 suggests that, because of the probability of vesicular emission, the presynaptic vesicular grids are ideally fitted to be the targets for the non-material mental events such as the intention to carry out some movement. It is *not* proposed that the mental events initiate activity at a synapse by an excitatory action either on the presynaptic or postsynaptic elements of a synapse such as Figures 4.4 and 4.5. On the contrary, the hypothesis is that the mental events merely alter the probability of a vesicular emission that is triggered by a presynaptic impulse. This action of a mental event would be exerted on the paracrystalline presynaptic vesicular grid that acts in a global manner in controlling the probability of emission of one vesicle from its array of many embedded vesicles.

The first question that can be raised concerns the magnitude of the effect that could be produced by a probability wave of quantum mechanics. Is the mass of the synaptic vesicle so great that it lies outside the range of the uncertainty principle of Heisenberg? Margenau (1977, p. 384) adapts the usual uncertainty equation for this calculation of non-atomic situations:

$$\Delta x \Delta v \geq k/m, \quad \text{where} \quad k = 1.06 \times 10^{-27}\,\text{erg s}.$$

The mass (m) of a synaptic vesicle 40 nm in diameter (1 nm = 10^{-9} m) can be calculated to be 3×10^{-17} g. If the uncertainty of the position Δx of the vesicle in the presynaptic vesicular grid is taken to be 1 nm, then Δv, the uncertainty of the velocity, comes out at 3.5 nm in 1 ms, which is not far from the right order of magnitude. The presynaptic membrane (Figures 4.4 and 4.5) is about 5 nm across and the time of emission of a vesicle is many tenths of a millisecond (Katz and Miledi 1965).

However, this calculation assumes that the synaptic vesicle is freely moving, which is certainly not the case when it is embedded in the presynaptic vesicular grid (Figure 4.4). Since that grid is a paracrystalline structure, it could have special resonance relations with a mental influence operating analogously to a probability field. A valuable insight into the manner of operation of the presynaptic vesicular grid could come from the quantum mechanics of microcrystalline structures. As illustrated in Figures 4.4 and 4.5, the postulated mental influence would do no more than alter the probability of emission of a vesicle already in apposition.

It can be concluded that calculations on the basis of the Heisenberg uncertainty principle show that the probabilistic emission of a vesicle from the presynaptic vesicular grid could conceivably be modified by a mental intention acting analogously to a quantal probability field.

The second question raises the order of magnitude of the effect, which is merely a change in probability of emission of a single vesicle (Figure 4.5). This is many orders of magnitude too small for modifying the patterns of neuronal activity even in small areas of the brain. However, there are many thousands of similar boutons on a pyramidal cell of the cerebral cortex. *The hypothesis is that the probability field of the mental intention is widely distributed not only to the synapses on that neuron but also to the synapses of a multitude of other neurons with similar functions of the dendron* (see Chapter 6). The next section will treat this problem with special reference to the cerebral response to the mental intention to carry out a voluntary movement.

5.3 Amplification of the Postulated Action of a Mental Intention in Changing the Probability of Emission of a Synaptic Vesicle

Figure 5.1 shows the position of the supplementary motor area (SMA) of the left cerebral hemisphere in the medial part of the frontal cortex just anterior to the motor area of the hind limb and extending deep on the medial side. By a radio-xenon technique, Roland et al. (1980) recorded the regional blood flow (rCBF) over a cerebral hemisphere, there being an assembled pattern from 254 Geiger counters for recording the detailed spatial pattern of radio-emission following a brief injection of radio-xenon into the internal carotid artery. It is now established that any regional increase in rCBF is a reliable signal of an increased neuronal activity in that area. The subject was trained to make a complex pattern of finger–thumb movements for the full duration (45 s) of the Geiger counting. In Figure 5.2a there was a strong activation of the contralateral motor and sensory areas for the thumb and fingers, as would be expected, but there was just as strong an activation of the SMA, and that was bilateral. The primacy of the SMA is revealed in Figure 5.2b, when during the radio-xenon test the subject was making no movement but merely carrying out the learned motor task mentally. A highly significant (20%) increase in neuronal activation was restricted to the SMA on both sides and was nowhere else. The subject was at complete rest with eyes and ears closed. This rCBF increase is an index of an increase

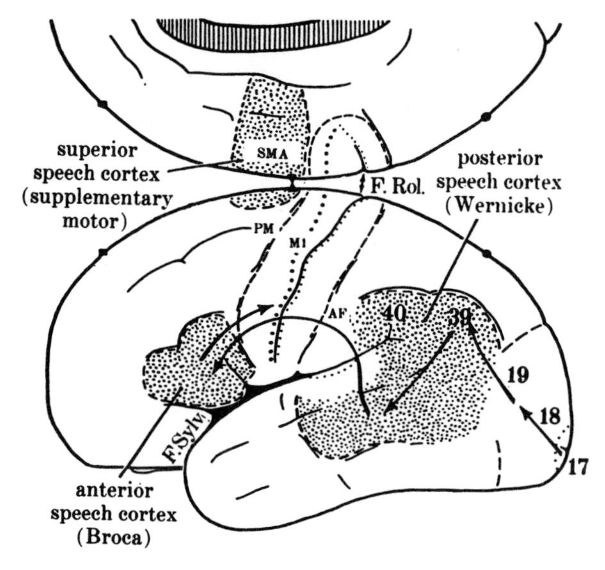

Figure 5.1 The left hemisphere from the lateral side with frontal lobe to the left. The medial side of the hemisphere is shown as if reflected upwards. F. Rol. is the fissure of Rolando, or the central fissure; F. Sylv. is the fissure of Sylvius. The primary motor cortex (M1) is shown in the precentral cortex just anterior to the central sulcus and extending deeply into it. Anterior to M1 is shown the premotor cortex (PM) with the supplementary motor area (SMA) largely on the medial side of the hemisphere (modified from Penfield and Roberts 1959).

in neuronal activity of the SMA under the influence of a mental intention by the subject. Evidently the mental intention was bringing into action an immense ensemble of neurons, which of course would be essential if it is to cause the desired movement.

By means of an implanted microelectrode it has been possible to study the responses of single SMA neurons of a monkey while it was carrying out a voluntary movement (Figure 5.3a, c; Brinkman and Porter 1979). There was an increase in the discharge rate of many neurons at about 50 ms before the discharge of motor cortical neurons that eventually would cause the willed movement as signalled by the electromyogram of Figure 5.3b (cf. Eccles 1982a, b). Ethical considerations preclude the carrying out of such an experiment on a human subject. However, the recording of electric and magnetic fields over the human scalp during repetitive voluntary movements (Deecke and Kornhuber 1978) also point to the neurons of the SMA as the site of strong activation by the mental intention.

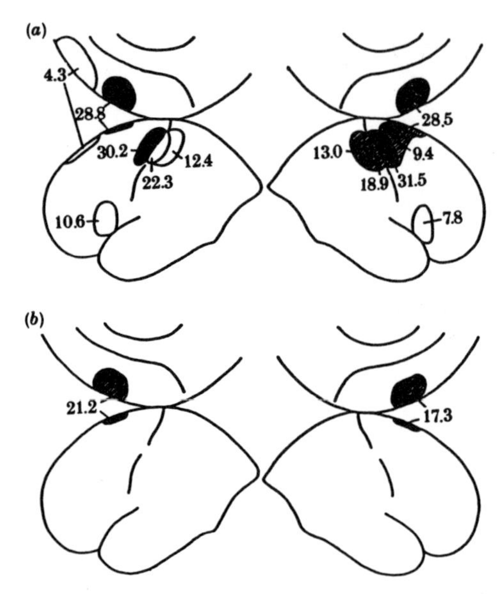

Figure 5.2 (**a**) The mean increase of the rCBF (%) during the motor-sequence test performed with the contralateral hand, corrected for diffuse increase of the blood flow. *Cross-hatched areas* have an increase of rCBF significant at the $p = 0.0005$ level. *Hatched areas* have an increase of rCBF significant at the $p = 0.005$ level. For other areas shown the rCBF increase is significant at the $p = 0.05$ level. *Left:* left hemisphere, five subjects. *Right:* right hemisphere, ten subjects. (**b**) The mean increase of rCBF (%) during internal programming of the motor-sequence test: values are corrected for diffuse increase of the blood flow. *Left:* left hemisphere, three subjects. *Right:* right hemisphere, five subjects (Roland et al. 1980).

According to the dualist-interactionist hypothesis (see Chapters 1.4 and 2.2) *the presynaptic vesicular grid provides the chance for the mental intention to change by choice the probability of its synaptic emission.* This would be happening over the whole ensemble of spine synapses that are activated at that time, probably even thousands, since there are about 10 000 on a single cortical pyramidal cell (Szentágothai 1978a). It would be expected that a mental influence analogous to a probability field would exert a global influence on the synapses of an appropriate neuron, modifying the probabilities of vesicular emission by incoming impulses.

So the *reliability* of mental intention is derived from integration of the *chance happenings* at the multitude of presynaptic vesicular grids on that neuron. In order to bring about some chosen movement, such as bending one's finger, the mental intention has to select the correct pyramidal cells for its action in modifying the probability of vesicular emissions. This

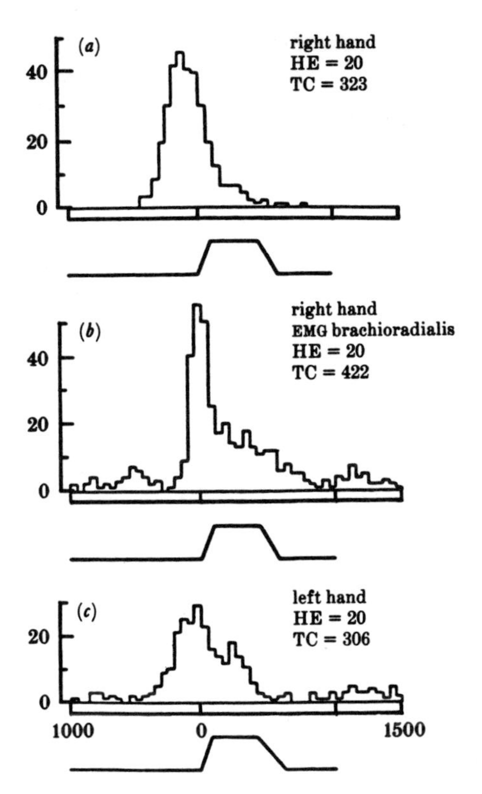

Figure 5.3 Illustration of the discharge patterns of a neuron associated with flexion of the elbow during the lever pull for both the right hand (**a**) and left hand (**c**). (**b**) is the periresponse time histogram demonstrating the EMG activity of a representative elbow flexor, m. brachioradialis, in the right arm during the same 20 pulls as those in (a), and shows that the neuron increased its discharge well before EMG activity increased. This was the case for the majority of neurons in which the discharge pattern could be compared with EMG changes (Brinkman and Porter 1979).

selection is accomplished according to the *learned inventory of SMA cells for a particular movement.* It can be effective only when there is a background synaptic barrage on those cells, because all it can do is to modify the probability of vesicular emission of the activated synapses. Tanji and Kurata (1982) have demonstrated the wide convergence of sensory inputs on to SMA cells. All such activated boutons can be regarded as presumptive sites for modification of the probability of vesicular release by the mental intention.

This may seem a clumsy method for initiating a movement, but it must be recognized that we do have the ability to carry out at will an immense range of movements, and this necessitates a most complex strategy of selection of SMA neurons from the enormous inventory of about 100 million pyramidal cells in perhaps 30 000 modules (cf. Chapter 6). All we experience is how mentally to initiate the skilled movement. It is essential that the mental intention causes the activation of packages of SMA neurons in correct time sequence for the different muscles concerned in the motor act, as has been demonstrated for SMA neurons by Brinkman and Porter (1979, 1983). In the simplest explanation these SMA cells project to the other cortical and subcortical areas in order to have the learned motor programmes incorporated into the eventual activation of the motor pyramidal cells with discharge down the pyramidal tract (Figure 4.7).

In summary it can be stated that it is sufficient for the dualist-interactionist hypothesis to be able to account for the ability of a non-material mental event to effect a changed probability of the vesicular emission from a single bouton on a cortical pyramidal cell. If that can occur for one, it could occur for a multitude of the boutons on that neuron, and all else follows in accord with the neuroscience of motor control. The closedness of World 1 has been opened and by a mental intention we genuinely are able to bring about movements at will.

5.4 The Action of Silent Thinking on the Cerebral Cortex

Figure 5.4a illustrates a remarkable finding of Roland (1981) that, when the human subject was attending to a finger on which a just-detectable touch stimulus was to be applied, there was an increase in the rCBF over the finger touch area of the postcentral gyrus of the cerebral cortex as well as in the midprefrontal area. These increases must have resulted from the mental attention because *no touch was applied during the recording*. Thus, Figure 5.4a is a demonstration that the mental act of attention can activate appropriate regions of the cerebral cortex. A similar finding occurs with attention to the lips in expectation of a touch, but of course the activated somatosensory area is now for the lips.

A related finding is that, when the subject was attending to simple counting or other arithmetical mental activities during complete relaxation with eyes and ears closed, there was an increased rCBF in many cortical areas, but not in the primary sensory or motor areas (Roland and Friberg 1985). As illustrated in Figure 5.4b for both the left and right hemispheres with

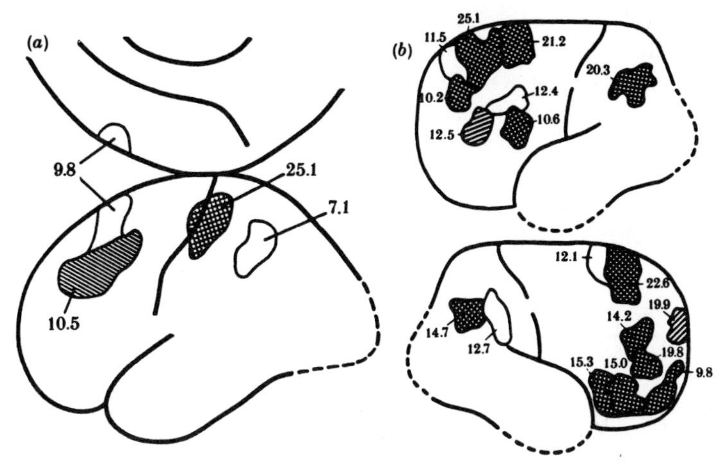

Figure 5.4 (**a**) The mean increase of rCBF (%) during pure selective somatosensory attention, that is, somatosensory latent imaging without peripheral stimulation. The size and location of each focus shown is the geometrical average of the individual focus. Each individual focus has been transferred to a brain map of standard dimensions with a proportional stereotactic system. The *cross-hatched areas* have an increase of rCBF significant at the $p = 0.0005$ level (Student's t test, one-sided significance level). For the *other areas* shown the rCBF increase is significant at the $p = 0.05$ level. Eight subjects (Roland 1981). (**b**) Mean increases of rCBF (%) and their average distribution in the cerebral cortex during silent arithmetic with successive subtraction of threes from 50. Left hemisphere, six subjects; right hemisphere, five subjects. *Cross-hatched areas* have rCBF increases significant at the $p = 0.005$ level. With *hatched areas* $p < 0.01$ and with *outlined areas* $p < 0.05$ (Roland and Friberg 1985).

the silent mental arithmetic of successive subtractions of 3 starting from 50, there was an increased rCBF in a medial strip of the frontal cortex anterior to the SMA and also in other areas of the prefrontal cortex on both sides, as well as in the supramarginal and angular gyri of both parietal lobes. The patterns are more complex than for the silent thinking of a motor movement in Figure 5.2b. Still more complex patterns were revealed with a memory sequence based on a nonsense word sequence and with the visual imagery of route-finding. (In Chapter 10.5 there is full reference to the effect of attention on the neocortex; see Figures 10.1 and 10.3.)

It can be predicted that the immense range of silent thinking of which we are capable will be found to initiate activity in such a wide variety of specific regions of the cerebral cortex that the greater part of the neocortex will be found to be under the mental influence of thinking (Ingvar 1985). Of course there is as yet no criterion for demonstrating a direct influence. The areas of direct activation can immediately influence other areas, as occurred with the SMA (Figure 5.2b) activating the motor cortex (Figure 5.2a).

The hypothesis that non-material mental events alter the probability of vesicular emission by presynaptic vesicular grids can account for all these influences of silent thinking, as described in principle in Chapter 9.

5.5 The Mind–Brain Problem

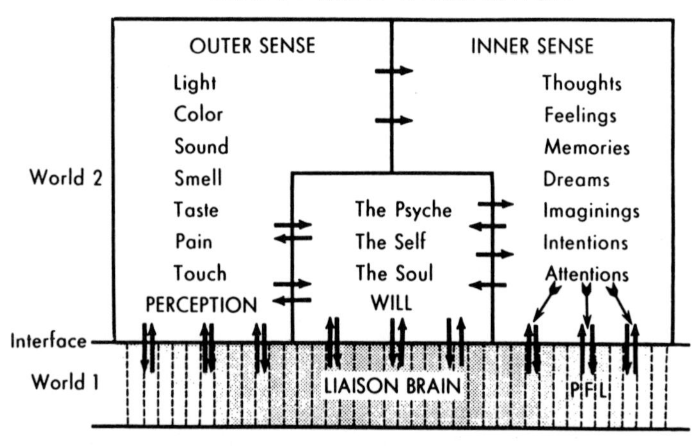

Figure 5.5 An information flow diagram for mind–brain interaction in the human brain. The three components of World 2 (outer sense, inner sense, and the ego, self, or soul) are drawn with their communications shown by arrows. Also shown are the lines of communication across the interface between World 1 and World 2, that is, from the liaison brain to and from these World 2 components. The liaison brain has the columnar arrangement indicated by the *vertical broken lines*. It must be imagined that the area of the liaison brain is enormous, with open or active modules numbering over a million, not just the two score depicted here. The supplementary motor area, SMA, is shown specially related to intentions of World 2, with the *three arrows* giving some suggestion of the potential specificity of action of the intention on the modules of the SMA, as discussed in the text. World 2 is shown above World 1, but this is a diagrammatic device without spatial significance. If World 2 is to be given any spatial location, it will be placed where it acts, which is shown by the *arrows* to be in the modules of the liaison brain.

In formulating more precisely the dualist hypothesis of mind–brain interaction, the initial statement is that the whole world of mental events (World 2) has an existence as autonomous as the world of matter–energy (World 1) (Figure 5.5). It can be mentioned that we only know of World 1 through sense organs. Sense organs provide the data by which we perceive and act and think and remember, and hence for all of human activity including science and technology. The present interactionist hypothesis does not relate to these ontological problems but merely to the mode of action of mental events on neural events, that is, to the nature of the downward arrows across the frontier in Figure 5.5. *The hypothesis is that the mental influence modifies the probability of vesicular emission from an activated bouton in a manner analogous to the probability fields of quantum mechanics.*

Since it is postulated that mental events can influence only those neural events engaged in the probability of quantal (vesicular) emission by presynaptic impulses, it would be predicted that the effectiveness of mental events would be reduced to zero when the presynaptic background was reduced to zero. Loss of consciousness would occur and would be irreversible unless there were revival to a considerable degree of the impulse discharges in the cerebral cortex. An example is 'vigil coma' that supervenes when injury to the midbrain turns off the reticular activating system (Hassler 1978; Eccles 1980). In fact the principal role of the reticular activating system may be to provide a background of excitatory impulses into the cerebral cortex with an immense array of probabilistic vesicle emissions that are targets for the quantal probabilistic fields of mental influence.

So we can assume that, in a global manner, the mental events achieve interaction with the neural events of spatio-temporal patterns of activity (Eccles 1982b) of the awake cerebral cortex. Even in one cortical module with its 4000 or so neurons there must be an on-going intense dynamic activity of unimaginable complexity. Although we know the outlines of the neuronal structure of a module (Szentágothai 1978a, 1983), there has as yet been only a very limited study of the physiology. All that we can surmise is that mental events acting as a field in the manner postulated by Margenau (1984) could effect changes in the spatio-temporal activity of a module by changing the probability of emission in many thousands of active synapses. There need be no violation of conservation laws.

One can ask how the monkey sets up the immense synaptic barrage that results in the neuronal firing in Figure 5.3a, c and that through the well-known complex pathways (Figure 4.7) results in the desired motor action. The only answer is that this performance is at the end of a long line of training sessions. Motor learning is essential for all skilled actions devolving from the cerebral cortex, and this is particularly true for human actions (Eccles 1986). *Memory of some kind is required* for all conscious experiences and actions.

A final consideration relates to the reverse arrows in Figure 5.5 from cerebral cortex to mind, as, for example, in perception to the left of the diagram. Is it possible that the discharge of a vesicle from a presynaptic vesicular grid can cause a mental event by a quantum probability wave in the reverse direction? Such vesicular emissions would occur in immense numbers in the perceptual areas of the cerebral cortex. Hence there could be an immense summation of 'unitary' mental events in order to surpass the threshold for a perception.

A general observation is that hitherto all hypotheses attempting to give some explanation of how conscious experiences derive from or relate to neural events concentrate on the extreme complexity of the neural events in the active cerebral cortex, as done by Feigl (1967) in the introduction. Sperry (1976) proposed that mental events are *holistic configurational properties of the brain process.* Mountcastle (1978) developed the concept of *distributed systems* which are

> composed of large numbers of modular elements linked together in echeloned parallel and serial arrangements,

and are thought to provide an objective mechanism of conscious awareness. Edelman (1978) suggested that

> the brain processes sensory signals and its own stored information upon this selective base in a phasic (cyclic) and reentrant manner that is capable of generating the necessary conditions for conscious states.

Szentágothai (1978b) suggested that

> *dynamic patterns offer 'superstructures'*

and might be helpful in giving a scientific explanation of the higher functions of the brain, including even consciousness. I (1982b) suggested that

> the mental influence is exerted on an extremely complex dynamic system of interacting neurons.

The extreme alternative to these *'nebular' hypotheses* is now proposed, namely that the essential locus of the action of the self on the brain is at individual microsites, the presynaptic vesicular grids of the boutons, each of which operates in a probabilistic manner in the release of a single vesicle in response to a presynaptic impulse. It is this probability that is assumed to be modified by the self acting analogously to a quantal probability field in the manner described in Chapter 9 and illustrated in Figures 9.5 and 10.2. The manner in which effective action at microsites becomes amplified by conventional neurocircuitry will be dependent on the complex circuits envisioned, for example, by Feigl (1967), Sperry (1976), Edelman (1978),

Mountcastle (1978), Szentágothai (1978b), and myself (1982b). *The microsite hypothesis* can be proposed as a tentative beginning of a scientific study of the reflective loop proposed by Creutzfeldt (1979) as opening up the independent symbolic world of the mind, which is the World 2 of Popper and myself (1977). In contrast to the 'nebular' hypotheses it offers a unique challenge to molecular neurobiology.

References to Chapters 4 and 5

Akert, K., Peper, K., and Sandri, C. (1975) Structural organization of motor end plate and central synapses, in *Cholinergic Mechanisms*, edited by E. G. Waser (Raven Press, New York), pp. 43–57.

Brinkman, C., and Porter, R. (1979) Supplementary motor area in the monkey: activity of neurons during performance of a learned motor task, *J. Neurophysiol.* **42**, 681–709.

Brinkman, C., and Porter, R. (1983) Supplementary motor area and premotor area of the monkey cerebral cortex: functional organization and activities of single neurons during performance of a learned movement, *Adv. Neurol.* **39**, 393–420.

Brown, A. G. (1981) *Organization in the Spinal Cord: The Anatomy and Physiology of Identified Neurones* (Springer, Berlin, Heidelberg).

Burke, R. E., Walmsley, B., and Hodgson, J. A. (1979) HRP anatomy of group Ia afferent contacts on alpha motoneurones, *Brain Res.* **160**, 347–352.

Creutzfeldt, O. D. (1979) Neurophysiological mechanisms and consciousness, in *Brain and Mind* (Ciba Foundation Series 69) (Elsevier–North Holland, Amsterdam), pp. 217–233.

Deecke, L., and Kornhuber, H. H. (1978) An electrical sign of participation of the mesial 'supplementary' motor cortex in human voluntary finger movement, *Brain Res.* **159**, 473–476.

DeLorenzo, R. J. (1981) The calmodulin hypothesis of neurotransmission, *Cell. Calcium* **2**, 365–385.

Eccles, J. C. (1980) *The Human Psyche* (Springer, Berlin, Heidelberg).

Eccles, J. C. (1982a) The initiation of voluntary movements by the supplementary motor area, *Arch. Psychiatr. Nervenkr.* **231**, 423–441.

Eccles, J. C. (1982b) How the self acts on the brain, *Psychoneuroendocrinol.* **7**, 271–283.

Eccles, J. C. (1986) Learning in the motor system, *Prog. Brain Res.* **64**, 3–18.

Eccles, J. C., Eccles R. M., and Lundberg, A. (1957) Synaptic actions on motoneurones in relation to the two components of the group I muscle afferent volley, *J. Physiol., London* **136**, 527–546.

Edelman, G. M. (1978) Group selection and phasic reentrant signalling: a theory of higher brain function, in *The Mindful Brain* (MIT Press, Cambridge MA), pp. 51–100.

Feigl, H. (1967) *The 'Mental' and the 'Physical'* (University of Minnesota Press, Minneapolis MN).

Gray, E. G. (1982) Rehabilitating the dendritic spine, *Trends Neurosci.* **5**, 5–6.

Hassler, R. (1978) Interaction of reticular activating system for vigilance and the truncothalamic and pallidal systems for directing awareness and attention under striatal control, in *Cerebral Correlates of Conscious Experience*, edited by P. A. Buser and A. Rougeul-Buser (Elsevier–North Holland, Amsterdam), pp. 110–129.

Hirst, G. D. S., Redman, S. J., and Wong, K. (1981) Post-tetanic potentiation and facilitation of synaptic potentials evoked in cat spinal motoneurones, *J. Physiol., London* **321**, 97–109.

Hubbard, J. I. (1970) Mechanism of transmitter release, *Prog. Biophys. Molec. Biol.* **21**, 33–124.

Ingvar, D. H. (1985) 'Memory of the future.' An essay on the temporal organization of conscious awareness, *Hum. Neurobiol.* **4**, 127–136.

Jack, J. J. B., Redman, S. J., and Wong, K. (1981a) The components of synaptic potentials evoked in cat spinal motoneurones by impulses in single group Ia afferents, *J. Physiol., London* **321**, 65–96.

Jack, J. J. B., Redman, S. J., and Wong, K. (1981b) Modifications to synaptic transmission at group Ia synapses on cat spinal motoneurones by 4-aminopyridine, *J. Physiol., London* **321**, 111–126.

Katz, B., and Miledi, R. (1965) The measurement of synaptic delay and time course of acetylcholine release at neuromuscular junction, *Proc. Roy. Soc. London B* **161**, 483–495.

Kelly, R. B., Deutsch, J. W., Carlson, S. S., and Wagner, J. A. (1979) Biochemistry of neurotransmitter release, *A. Rev. Neurosci.* **2**, 399–446.

Korn, H., Mallet, A., Triller, A., and Faber, D. S. (1982) Transmission at a central inhibitory synapse. II. Quantal description of release, with a physical correlate for binomial n, *J. Neurophysiol.* **48**, 679–707.

Korn, H., and Faber, D. S. (1985) Regulation and significance of probabilistic release mechanisms at central synapses, in *New Insights into Synaptic Function*, edited by G. M. Edelman, W. E. Gall, and W. M. Cowan (Neurosciences Research Foundation/Wiley, New York), pp. 57–108.

Margenau, H. (1984) *The Miracle of Existence* (Ox Bow, Woodbridge, CT).

Mendell, L. M., and Henneman, E. (1971) Terminals of single Ia fibers: location, density and distribution within a pool of 300 homogeneous motoneurons, *J. Neurophysiol.* **34**, 171–187.

Mountcastle, V. B. (1978) An organizing principle for cerebral function: the unit module and the distributed system, in *The Mindful Brain* (MIT Press, Cambridge MA), pp. 7–50.

Penfield, W., and Robert, L. (1959) *Speech and Brain Mechanisms* (Princeton University Press, Princeton NJ).

Popper, K. R., and Eccles, J. C. (1977) *The Self and Its Brain* (Springer, Berlin, Heidelberg).

Redman, S. J. (1980) Mechanisms of transmitting release at Ia afferent terminations, *Adv. Physiolo. Sci.* **1**, 93–100. Also in *Regulatory Functions of the CNS. Principles of Motion and Organization*, edited by J. Szentágothai, M. Palkovits, and J. Hamori (Pergamon, Oxford).

Redman, S., and Walmsley, B. (1983a) The time course of synaptic potentials evoked in cat spinal motoneurones at identified group Ia synapses, *J. Neurophysiol.* **343**, 117–133.

Redman, S., and Walmsley, B. (1983b) Amplitude fluctuations in synaptic potentials evoked in cat spinal motoneurones at identified group Ia synapses, *J. Neurophysiol.* **343**, 135–145.

Roland, P. E. (1981) Somatotopical tuning of postcentral gyrus during focal attention in man. A regional cerebral blood flow study, *J. Neurophysiol.* **46**, 744–754.

Roland, P. E., and Friberg, L. (1985) Localization in cortical areas activated by thinking, *J. Neurophysiol.* **53**, 1219–1243.

Roland, P. E., Larsen, B., Lassen, N. A., and Skinhøj, E. (1989) Supplementary motor area and other cortical areas in organization of voluntary movements in man, *J. Neurophysiol.* **43**, 118–136.

Sperry, R. W. (1976) Mental phenomena as causal determinants in brain function, in *Consciousness of the Brain*, edited by G. G. Globus, G. Maxwell, and I. Savodnik (Plenum, New York), pp. 163–177.

Szentágothai, J. (1978a) The neuron network of the cerebral cortex. A functional interpretation, *Proc. Roy. Soc. London B* **201**, 219–248.
Szentágothai, J. (1978b) The local neuronal apparatus of the cerebral cortex, in *Cerebral Correlates of Conscious Experience*, edited by P. Buser and A. Rongeul-Buser (Elsevier, Amsterdam), pp. 131–138.
Szentágothai, J. (1983) The modular architectonic principle of neural centers, *Rev. Physiol. Biochem. Pharmacol.* **98**, 11–61.
Tanji, J., and Kurata, K. (1982) Comparison of movement-related activity in two cortical motor areas of primates, *J. Neurophysiol.* **48**, 633–653.
Triller, A., and Korn, H. (1982) Transmission at a central inhibitory synapse. III. Ultrastructure of physiologically identified and stained terminals, *J. Neurophysiol.* **48**, 708–736.

6 A Unitary Hypothesis of Mind–Brain Interaction in the Cerebral Cortex

6.1 Introduction

A large part of this chapter is devoted to an attempt to discover the anatomical structure of the neocortex that relates to the self–brain problem. There is firstly the beautiful diagram by Szentágothai (1975) of the neocortex with concentration on the pyramidal cell structure (Figures 6.5 and 7.1a). The important development was provided by the elegant studies by Fleischhauer of Bonn and Peters of Boston and their pupils. The bundling of the apical dendrites of pyramidal cells as they ascend to lamina I was revealed by precise histological studies with microphotographs (Figures 6.6–6.7) and is illustrated in the drawings of Figures 6.8, 6.10, and 8.2b. The bundles of dendrites are called dendrons and there are about 100 apical dendrites in each dendron.

The detailed structure of a spine synapse is illustrated in Figure 7.1b. There are over 100 000 spine synapses on the bundled apical dendrites of a dendron, which is thus an enormous receiving structure for excitatory synapses. Figure 7.2a illustrates imperfectly the clustering of synaptic boutons (Figure 7.1b) along the apical dendrite of a lamina V pyramidal cell and its branches. About 200 are drawn, but the actual number is over 5 000 for a lamina V pyramidal cell.

As originally expressed, the microsite hypothesis (Eccles 1986 and Chapters 4 and 5) was deficient in that it did not define precisely the mental events that were assumed to be acting on the neural events. They had a rather nebulous character. A radical development is now necessary in order to extend the microsite hypothesis to perception and to the whole range of subjective experiences in the World 2 of Figure 6.1. The new hypothesis is that all mental events and experiences, in fact the entire aggregate of the outer and inner senses of World 2 (Figure 6.1) are a composite of elemental or unitary mental events, which we may call *psychons*. It is further proposed that each of these psychons is reciprocally linked in some unique manner to its dendron (Figure 6.10). The dendron is a fixed anatomical structure except for the synaptic plasticity of learning, but functionally there are great variations in intensity of action according to neural inputs. It is similar functionally to the linked psychon, which can be at all levels of mental intensity from zero to a maximum functional linkage with its dendron. Psychons are not perceptual paths to experiences of World 2 (Figure 6.1). They *are* the experiences in all their diversity and uniqueness.

The self–brain interaction can now be considered on the basis of the unitary interaction of a psychon with its dendron. This has been diagrammed in Figure 6.10. The lamina V dendrons are shown in accord with the experimental evidence of Figures 6.6 and 6.7. Superimposed on each of these three dendrons are three psychons, each with its unique psychic character (indicated by squares, open squares, and solid circles) and each embracing the whole dendron. No doubt the congruity is idealized in the diagram, and of course there are multitudes of closely related psychons that could be represented similarly by squares, open squares, and solid circles and that are in unitary relation with similar dendrons (Figure 6.11). The three different psychons give some insight into the complexity of patterned relation between dendrons and psychons. There could be thousands of types of psychons, each with a matching type of dendron, the grand total being about forty million psychons for an estimated forty million dendrons of the human brain.

The unitary hypothesis transforms the manner of operation of the intention. If, for example, the *psychon* for a specific experience, an intention, is represented by the pattern of open squares on the central dendron of Figure 6.10, it can be seen that the intention acts on the whole dendron with its assembled pyramidal dendrites and their synapses, which could number up to 100 000. So the mental intention would have a large global operation on that dendron. On the unitary hypothesis the psychon would of course operate at each presynaptic vesicular grid (PVG) of its dendron in selecting, by means of the quantal probability field, a vesicle for exocytosis. However, collectively there could be tens of thousands of such PVG sites on one dendron, so great amplification is ensured by the unitary operation of the linked psychons and dendrons. One has to recognize that in a lifetime of learning the intention to carry out a particular movement would be channelled largely to those particular psychons that are linked to those dendrons of the neocortex (the SMA) (Figures 5.1, 5.2, and 5.5) that are appropriate for bringing about the required action.

Before these latest dendron–psychon developments, there was the problem of the order of magnitude of a mental intention that was acting by selection for exocytosis of a single vesicle. A great amplification was essential. However, in the dendron–psychon hypothesis a psychon has as its field for the exocytotic selection the 100 000 spine synapses on its dendron (Figure 6.10), which gives a possible amplification of its action by several orders of magnitude.

There was much discussion about the proposed origins of consciousness, as partly reviewed in Chapter 3 for the many authors, but all were inadequate since they paid little or no attention to the microstructure of the neocortex. Beck has already opened the way to utilizing the probabilities of exocytosis whereby mental events could effectively act on the brain (Chapter 9).

This chapter represents a much further development of a theoretical paper (Eccles 1986). First, it has been possible to identify the basic receptive units of the cerebral cortex, the dendrons, and to incorporate them into the brain–mind theory developed on the basis of quantum mechanics. Second, it is proposed that the whole mental world is composed of units, psychons, and that mind–brain interaction occurs between the closely linked units, each dendron with its psychon. Third, from these fundamental unitary concepts there will be developed a theory of perception from dendron activity to psychon experience, also on the basis of quantum mechanics.

6.2 Mind–Brain Interaction

It should be recognized that there has been a philosophical revolution since the days of Ryle (1949) and the behaviourists, who denied any scientific meaning to the philosophical concepts of consciousness and to the experiences of self-consciousness (Searle 1984), even by the materialists (Armstrong 1981; Dennett 1969, 1978; Hebb 1980). However, they feel that their materialist beliefs are not threatened thereby because they regard mental events as existing in an enigmatic sort of identity with neural events at higher levels of the brain, presumably in the cerebral cortex (Feigl 1967) (see Chapters 1 and 3.4). This strange postulate of identity is never explained, but it is believed that it will be resolved when we have a more complete scientific understanding of the brain, perhaps in hundreds of years; hence this belief is ironically termed promissory materialism (Popper and Eccles 1977).

I shall return to the identity theory after I present an alternative view of the brain–mind problem, namely dualist-interactionism (Popper and Eccles 1977), as shown in Figures 5.5 and 6.1. The essential feature is dualism. The whole world of conscious experiences, or mind, is labelled World 2 and is sharply separated from the brain in the materialist World 1 by an interface. For diagrammatic convenience, World 2 is drawn above the liaison brain in World 1, but actually it would be within the cortex, as shown by the origin

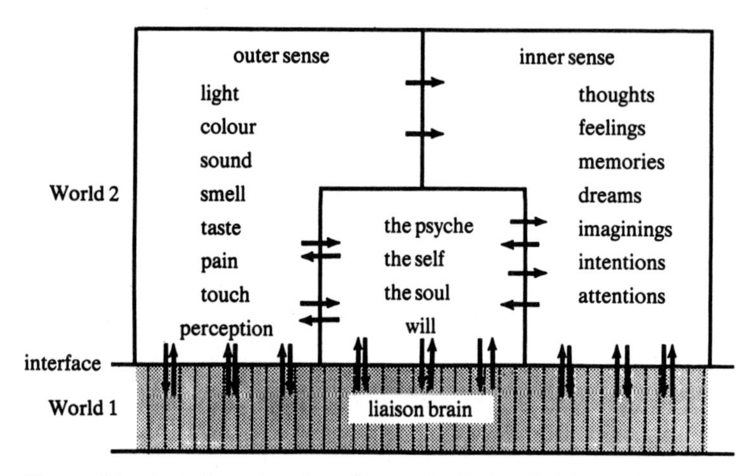

Figure 6.1 An information flow diagram for brain–mind interaction. The three components of World 2—outer sense, inner sense, and the psyche, or self—are shown with their connectivities. Also shown by *reciprocal arrows* are the lines of communication across the interface between World 1 and World 2, i.e. from the liaison brain to and from these World 2 components. The liaison brain has the columnar arrangement of its dendrons, which number about forty million.

and termination of the reciprocal arrows, which signify the interaction across the interface between the two Worlds. This diagram will develop more meaning later in the book.

6.3 The Neuronal Composition of the Cerebral Cortex

In Figure 6.2 the left cerebral hemisphere is shown in position in the head with some functional areas defined. The essential component is the *cerebral cortex*, which is a cellular layer about 3 mm thick covering the whole cortex even in its deep foldings so that it has an area of about 2 500 cm^2 for both hemispheres. The cerebral cortex is created by densely packed nerve cells with their associated nerve fibres, there being about 40 000 mm^{-2}, which would give a total of ten thousand million for the whole cortex. Figure 6.3a illustrates some nerve cells, which were selected from the tangled mass of a Golgi staining and drawn by the great Spanish neuroanatomist Ramón y Cajal. Of special importance are the pyramidal cells E, D, C, B, each with its apical dendrite ascending towards the surface and its axon descending

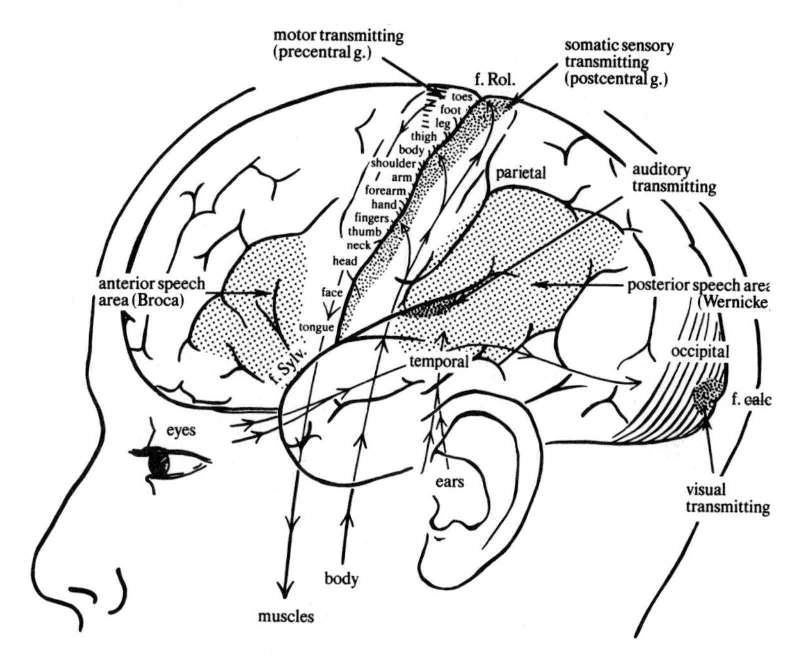

Figure 6.2 The motor and sensory transmitting areas of the cerebral cortex. The approximate map of the motor transmitting areas is shown in the precentral gyrus; the somatic sensory receiving areas are in a similar map in the postcentral gyrus. Actually, the toes, foot, and leg should be represented over the top on the medial surface. Other primary sensory areas shown are the visual and auditory, but they are largely in areas screened from this lateral view. Also shown are the speech areas of Broca and Wernicke.

to leave the cortex. At least 60% of cortical cells are pyramidal cells. Figure 6.3b shows three pyramidal cells, one in lamina V and two deep in lamina III.

The most important features of pyramidal cells are the small protuberances or spines that can already be seen in the dendrites in Figure 6.3a but are more defined in Figure 6.3b, where they are seen to be involved in close contacts with nerve fibres, making the spine synapses, which are the principal means of communication between nerve cells in the cortex. A wide variety of spine synapses is drawn in Figure 6.4a on a pyramidal cell of a special part of the cerebral cortex, the hippocampus. Finally Figure 6.4b shows an apical dendrite from the mouse visual cortex thickly encrusted by spines which can be counted, there being about one per micrometre and

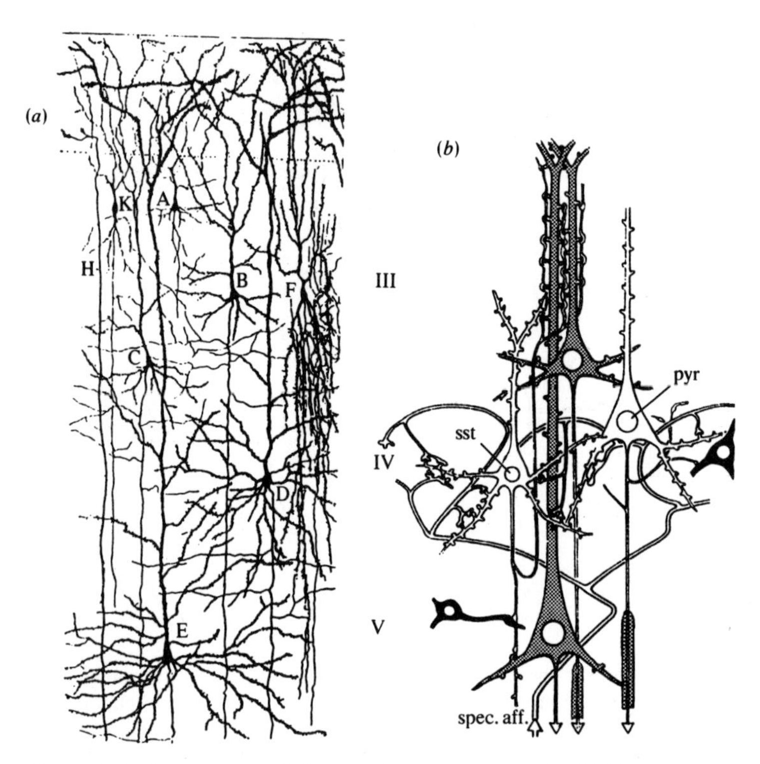

Figure 6.3 Neurons and their synaptic connections. (**a**) Eight neurons from a Golgi preparation of the three superficial layers of frontal cortex from a month-old child. Small (B, C) and medium (D, E) pyramidal cells are shown with their profuse dendrites covered with spines. Also shown are three other cells (A, F, K), which are in the general category of Golgi type II with their localized axonal distributions (Ramón y Cajal 1911). (**b**) The direct excitatory neuron circuit of the specific (sensory) afferents (spec. aff.). Both spiny stellate (sst) with ascending main axon and apical dendrites of both lamina III and V pyramidal cells (*stippled*) are probably the main targets (Szentágothai 1979).

hence at least 5 000 for each lamina V apical dendrite with its side branches and tufted ending (Figure 8.2).

In Figure 6.5 is illustrated the universally accepted six laminae of the cerebral cortex with two large pyramidal cells in lamina V, three in lamina III, and one (labelled *Sp*) in lamina II. There are many other pyramidal cells shown in shadowy outline and also several non-pyramidal cells, particularly in lamina IV. It will be noted that the pyramidal apical dendrites finish in a tuft-like branching in lamina I.

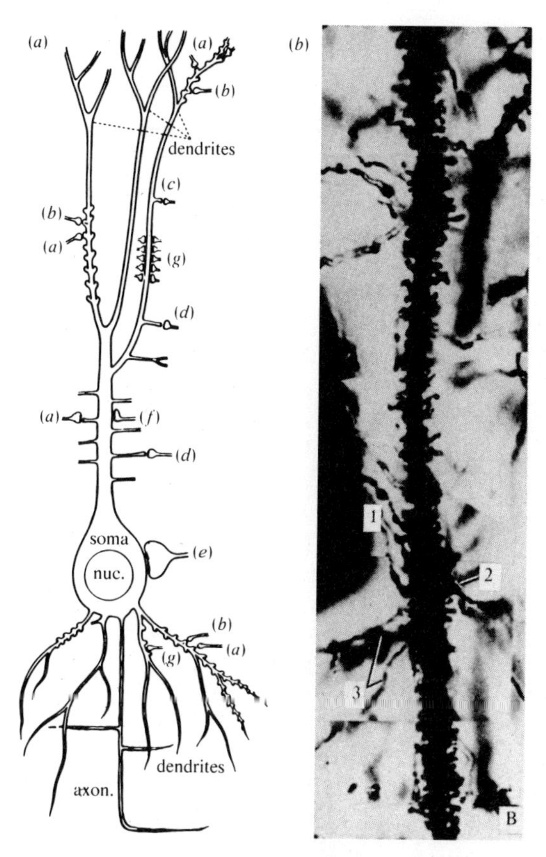

Figure 6.4 (**a**) A drawing of a hippocampal pyramidal cell to illustrate the diversity of synaptic endings on the different zones of the apical and basal dendrites and the inhibitory synaptic endings on the soma (Hamlyn 1963). (**b**) Mosaic micrographs of an apical dendrite of a pyramidal cell in the mouse area striata (Valverde 1968).

6.4 The Basic Receptive Unit of the Cerebral Cortex: The Dendron

Despite the intense microscopic study of the cerebral cortex, the basic neuronal assembly was not appreciated until Fleischhauer (Fleischhauer et al. 1972) and Feldman and Peters (1974) recognized the tendency of the pyramidal apical dendrites to assemble together in small bundles or clusters on their route to lamina 1. Since that time they and their associates have convincingly demonstrated this vertical assemblage of apical dendrites from

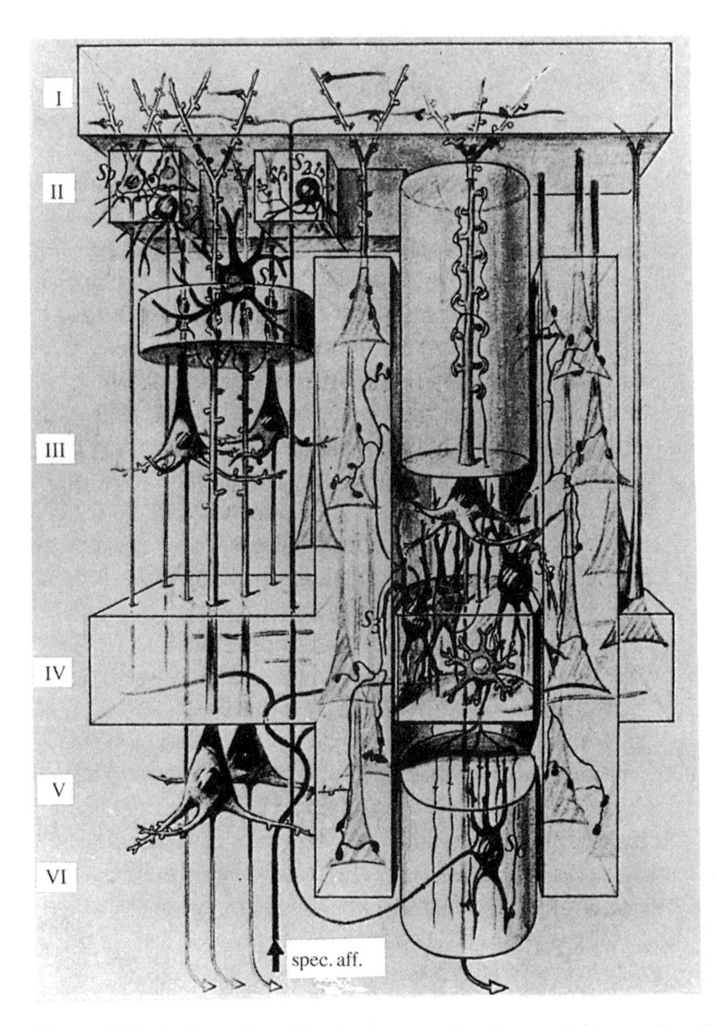

Figure 6.5 A three-dimensional construct showing cortical neurons of various types. There are two pyramidal cells in lamina V and three in lamina III, one being shown in detail in a column to the right (Szentágothai 1975).

pyramidal cells in lamina V that were joined in more superficial laminae by the apical dendrites of pyramidal III and II cells (Schmolke and Fleischhauer 1984; Feldman 1984; Schmolke 1987; Peters and Kara 1987).

In the photomicrograph of Figure 6.6a a Golgi staining of the rat visual cortex reveals large and medium pyramidal cells of lamina V, six being labelled: 1, 2 large and 3, 4, 5, 6 medium. Each apical dendrite ascends directly through the laminae. The bundles or clusters are not well recognized in Figure 6.6a because only a fraction of the pyramidal cells is stained by the Golgi method. However, the apical dendrites of pyramidal cells 1, 2, and 3 can be seen in close apposition. In the transverse section of Figure 6.6b at the level of the broken line in lamina IV of Figure 6.6a, there are many groups of transversely sectioned dendrites.

In a similar preparation (Figure 6.7a) many discrete clusters can be seen, one being indicated by the three open arrows. Figure 6.7b shows an electron microscopic (EM) picture of a transversely sectioned apical dendrite with the origins of three spines, the most prominent being labelled with an arrow and S.

Figures 6.6–6.8 enable the visualization of the bundles or clusters of apical dendrites. It is most clearly shown in transverse sections in lamina IV (Figures 6.6b and 6.7a) before there is the large addition from laminae III and II pyramidal cells (Peters and Kara 1987). As a consequence of this large accretion to the bundle there may be partial confluence of bundles in lamina II. The perspective drawing (Figure 6.8) by Schmolke (1987) gives an excellent picture of the growth of a bundle (Figure 8.2) from lamina V to laminae III–II. Three bundles are shown in the transverse section at the top. Figure 6.8 is simplified by neglecting the 20–30% of apical dendrites that do not participate in the bundles (Peters and Kara 1987). The average number of apical dendrites participating is about 8 large and 30 medium lamina V pyramidal cells with accretion to a total of 70–100 in lamina II before ending in the apical tufts of lamina I (Figures 6.5, 6.8, and 8.2).

There are many small pyramidal cells in laminae IV and VI, but their apical dendrites have a quite different course, ascending only up to lamina III before tufting. They do not join the bundles. Thus about one half of the neurons of the cerebral cortex do not participate in the bundles or clusters.

Despite this partial disarray there is agreement by Peters and Fleischhauer and their associates that the apical bundles or clusters diagrammatically shown in Figures 6.8 and 8.2 are the basic anatomical units of the cerebral cortex (Schmolke and Fleischhauer 1984; Peters and Kara 1987). They are observed in all areas of the cortex that have been investigated and in all mammals including humans.

Hitherto there has been no satisfactory suggestion of the functional role of these anatomical units. It is now proposed that they are the cortical units for reception, which would give them a preeminent role; hence it is desirable to name them. Since they are composed essentially of dendrites, the name *dendron* is proposed. In the nineteenth century, dendron was an alternative term for dendrites, but it has fallen into disuse in this century. For example,

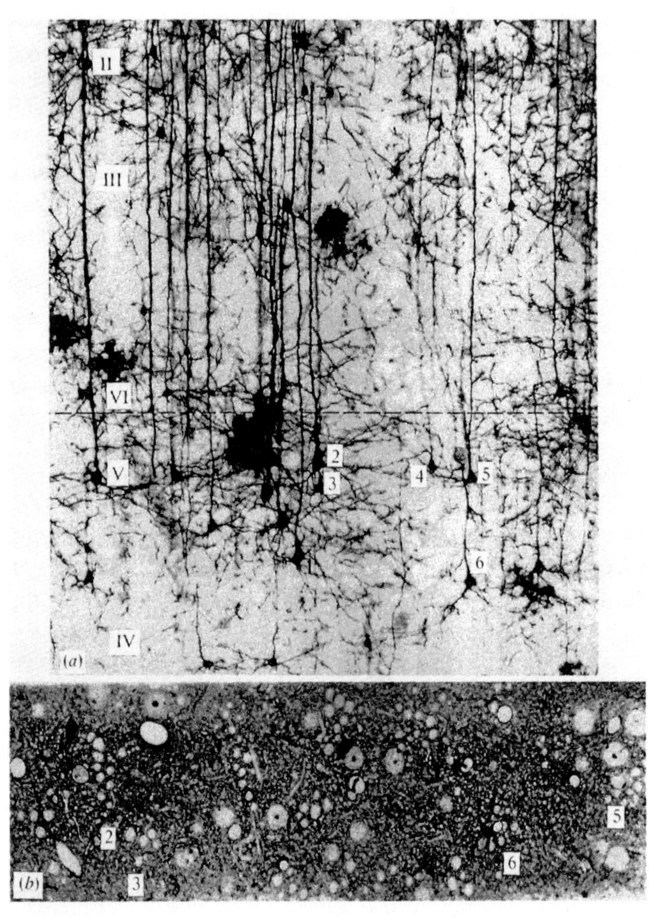

Figure 6.6 (**a**) A Golgi preparation of the rat visual cortex (area 17) in a vertical section through laminae II–VI to show apical dendrites of lamina V pyramidal cells projecting through laminae IV, III, and II. (**b**) A tangential section of lamina IV at the level indicated by the *broken line* in (a). The arrangement of the apical dendrites (*small open circles*) in clusters can be well seen (Peters and Kara 1987).

after 1900 it was not used by Ramón y Cajal or by Sherrington. As the proposal is that the dendron is a fundamental neural unit of the cerebral cortex, the ending of 'on' appropriately links it with the units of physics.

Approximate values can be given for the synaptic connectivity of a dendron. The input would be largely by the spine synapses (Figures 6.3b and 6.4a, b), which could be over 5 000 on a large dendrite, a lamina V apical

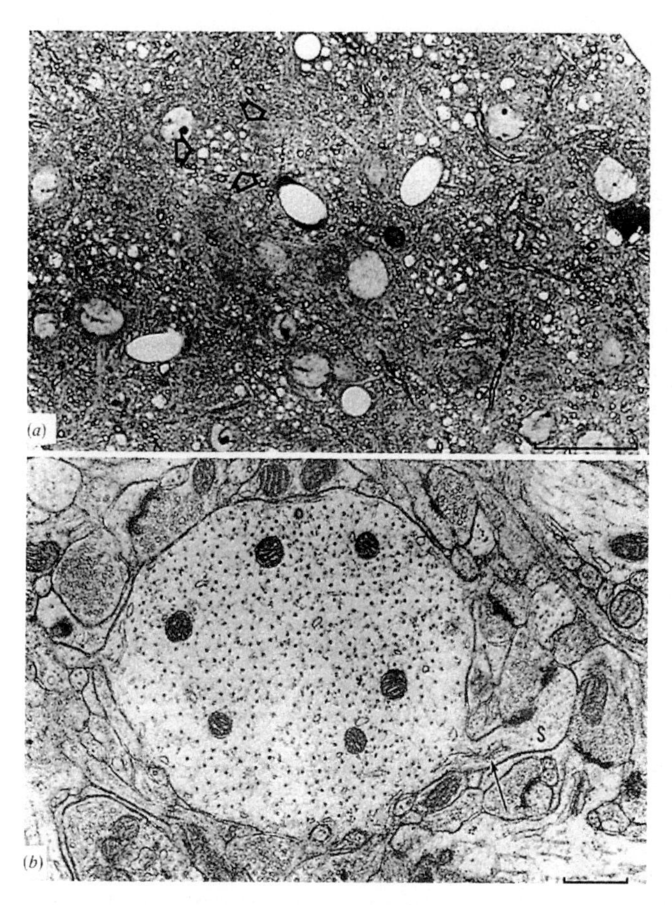

Figure 6.7 (**a**) Apical dendritic clusters as visualized in a tangentially oriented section at the level of layer V of rat visual cortex. One cluster is indicated by *open arrows*. *Calibration line*: 25 μm. (**b**) The large profile is a transversely sectioned apical dendrite within a dendritic cluster, as visualized in a tangential section through layer IV of rat visual cortex. At S, a dendritic spine extending from the apical dendritic shaft forms an asymmetric synapse with an axon terminal. *Bar*: 0.5 μm. (Feldman 1984.)

dendrite with its lateral branches (Figure 7.2a) and its terminal tuft, but more usually it would be under 2 000. If there are 70–100 apical dendrites in a dendron, the total number of spine synapses would be over 100 000. In addition to the apical dendrite there are also numerous synaptic inputs to

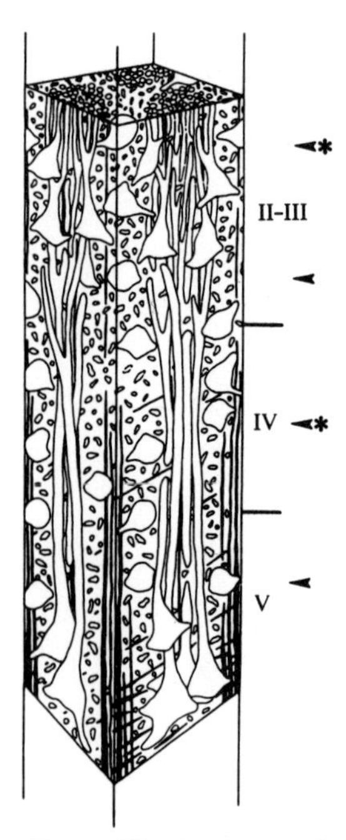

Figure 6.8 A stereoscopic diagram of a portion of rabbit visual cortex comprising laminae II–III, lamina IV, and the upper half of lamina V. The distribution of dendrites (*white*), myelinated axons (*black*), and perykarya (*white*) is shown (Schmolke 1987).

the basal dendrites and soma, as partly indicated in Figure 6.3b, but these are mostly inhibitory.

The synaptic output of a dendron is enormous. The axons of the pyramidal cells would mostly be distributed with modular transmission to the ipsilateral and contralateral cerebral hemispheres, as illustrated schematically in Figure 6.9 (Goldman and Nauta 1977; Szentágothai 1978). It is important to recognize that, as a basic anatomical unit of the cerebral cortex, the dendron has about 200 neurons in its region. The modules are transmission units defined by the cortico-cortical connectivities of the axons of the pyramidal cells (Figure 6.9) and each would contain about 4 000 neurons, which is about 20 times larger than the dendrons. There are about 200 dendrons and 10 modules per square millimetre of the cerebral cortex and about 40 million dendrons for the whole cortex.

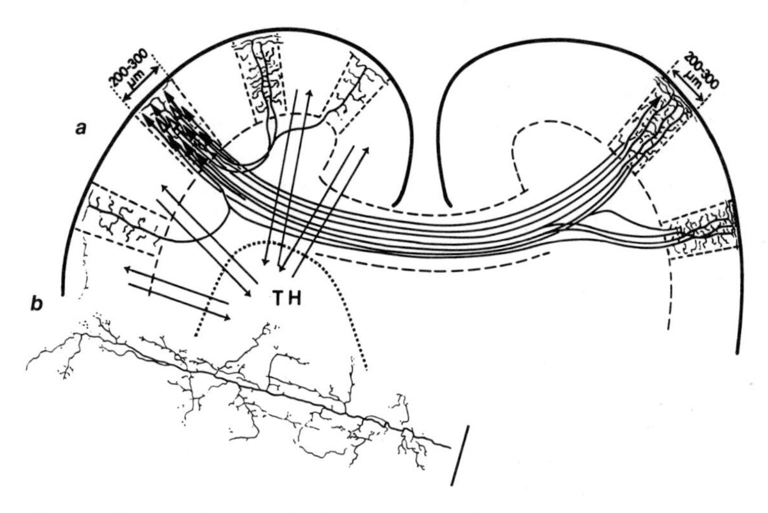

Figure 6.9 (**a**) The general principle of cortico-cortical connectivity shown diagramatically in a non-convoluted brain. The connections are established in highly specific patterns between vertical columns with a diameter of 200–300 μm in both hemispheres; TH = thalamus. (**b**) A Golgi-stained branching of a single cortico-cortical afferent, oriented in relation to the module with a single afferent in (a), but at several times higher magnification. (Szentágothai 1978.) *Scale bar*: 100 μm.

6.5 Mental Influences on the Brain

When one is at absolute rest in a darkened silent room, it is possible to engage in some specific thinking. For example, one can concentrate attention to a finger tip in order to detect a minimal touch that is expected. This attention causes neural activity in rather large areas of the brain, as is revealed by the rCBF technique (regional cerebral blood flow) (Roland 1981). In this technique, radio-xenon (^{133}Xe) is injected into the internal carotid artery through a cannula that has been inserted for a clinical investigation. A battery of 254 Geiger counters is mounted in a helmet applied over one side of the scalp. The brief injection causes a pattern of increased radioactivity, as observed by the counts of the Geiger assemblage. This increase in counts signals an increased blood flow, which in turn gives a quantitative measure of the subjacent cortical activity. The counts are done in the control resting situation and then during the chosen mental task, which for Figure 5.4a was concentrated attention on the finger tip in anticipation of a just recognizable touch.

There was an increase in the rCBF over the finger-touch area of the postcentral gyrus of the cerebral cortex (Figure 5.4a) as well as in the midprefrontal and parietal areas. These increases must have resulted from the mental attention, because no touch was applied during the recording. Thus Figure 5.4 is a clear demonstration that the mental act of attention can activate appropriate regions of the cerebral cortex. A similar finding occurs with attention to the lips in expectation of a touch, but of course the activated somatosensory area is now that for the lips. A large prefrontal area was also activated during the anticipation. Each of these areas is an expanse of several square centimetres and would include tens of thousands of dendrons. We can speculate on the mental attention and ask if it has a fine granular composition matching the tens of thousands of dendrons on which it acts.

A complementary investigation was on the mental intention for carrying out a complex learnt series of movements, the motor sequence tests (Figure 5.2a; Roland et al. 1980). When the subject was mentally rehearsing the movement sequence without carrying out any movement, there was a large mental activation of the supplementary motor area on both sides (Figure 5.2b). Again we can ask if the mental intention had a fine grain, being a composite of mental units matching the dendrons on which it was acting.

Even larger areas of cerebral cortex are activated by complex thinking procedures (Roland and Friberg 1985) such as the successive subtractions of threes from 50 (Figure 5.4b), or of the *imagined* experiences of walking along a well-known street, the subject being recumbent in a dark, quiet room (Figure 10.1c).

It has to be recognized that studies of the cerebral cortex lead on to the attempt to discover the mental events which interact with the dendrons in both attention and intention as shown by the reciprocal arrows across the interface in Figure 6.1. At present the experimental evidence (see Figures 5.4 and 10.1) is adequate only for establishing that mental intentions and attentions can indeed excite the dendron, but the observed actions are massive, tens of thousands of dendrons, presumably because of the operation of multitudes of mental events.

6.6 The Unitary Linkage of Mental Units to Dendrons of the Neocortex: The Psychon[1] Hypothesis

As originally expressed, the microsite hypothesis (Eccles 1986, p. 426) was deficient in that it did not precisely define the mental events that were assumed to be acting on the neural events. They had a rather nebulous character. A radical development is now necessary in order to extend the microsite hypothesis to perception and to the whole range of subjective experiences in the World 2 of Figure 6.1. The new hypothesis is that all mental events and experiences, in fact the entire aggregate of the outer and inner senses of World 2 (Figure 6.1), are a composite of elemental or unitary mental events, which we may call psychons. It is further proposed that each of these psychons is reciprocally linked in some unique manner to its dendron. The dendron is a fixed anatomical structure except for the synaptic plasticity of learning, but functionally there are great variations in intensity of action according to neural inputs. It is similar functionally to the linked psychon, which can be at all levels of mental intensity from zero to a maximum functional linkage with its dendron. Psychons are not perceptual paths to experiences of World 2 (Figure 6.1). They *are* the experiences in all their diversity and uniqueness.

The linkage has been crudely indicated in Figure 6.1 by the reciprocal arrows across the interface and now is more precisely drawn in Figure 6.10 in the manner of Figure 6.8. The lamina V apical dendrites are shown for three dendrons in accordance with the experimental evidence of Figures 6.6, 6.7, and 8.2. There has even been shown the small proportion of apical dendrites of lamina V that wander and do not join the dendron (Peters and Kara 1987). Superimposed on each of these three dendrons are three psychons, each with it unique psychic character (indicated by squares, open squares, and solid circles) and each embracing the whole dendron. No doubt the congruity is idealized in the diagram, and of course there are multitudes of closely related psychons that could be represented similarly by squares, open squares, and solid circles and that are in unitary relation with similar dendrons. The three different psychons give some insight into

[1] Professor Garrido has drawn my attention to the fact that M. Bunge has used the word 'psychon' in his book *The Mind–Body Problem* (Pergamon, 1980). Its derivation is described on page 37. However, it is a mistake to derive the word 'psychon' (from the Greek word ψυχή—'soul' or 'spirit'—(*Oxford English Dictionary*) for every plastic neural system (Bunge 1980, p. 56), which is a purely materialist concept. What Bunge needs is a word derived from the Greek word πλαστικός ('that may be moulded'), which would be 'plaston'. So Bunge should substitute 'plaston' for 'psychon' and leave 'psychon' for the purely mental usage proposed here.

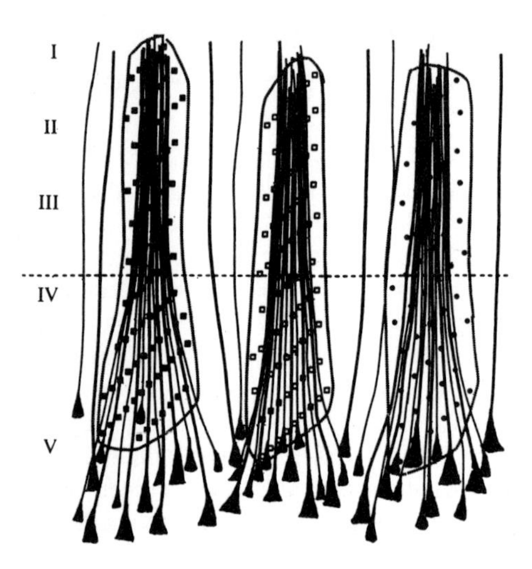

Figure 6.10 Drawings of three dendrons showing the manner in which the apical dendrites of large and medium pyramidal cells bunch together in lamina IV and more superficially, so forming a neural unit. A small proportion of apical dendrites do not join the bundles. The apical dendrites are shown terminating in lamina I. This termination is in tufts that are not shown. The other feature of the diagram is the superposition on each neural unit or dendron of a mental unit or psychon that has a characteristic marking (*solid squares*, *open squares*, *solid circles*). Each dendron is linked with a psychon giving its own characteristic unitary experience.

the complexity of patterned relation between dendrons and psychons. There could be thousands of types of psychons, each with a matching type of dendron, the grand total being about forty million psychons for the forty million dendrons.

Diagrammatically, when viewed from the cortical surface, each dendron can be illustrated as a circle in close relation with other dendrons, there being about 200 mm^{-2} (Figure 6.11). In this pattern of 64 dendrons the associated psychons are drawn with identification by the same three types of symbols as in Figure 6.10, to give some indication of the patterns of the various dendron–psychon units. Each such ensemble has its own psychic property, for example solid squares for a red light, solid circles for a touch, and open squares for an intention of some specific movement.

It may seem that in this intimate linkage of dendrons and psychons the new unitary hypothesis of dualist-interactionism is merely a further refinement of the materialist identity hypothesis (Feigl 1967) (see Chapter 6.2). This is a mistake. Independence of existence is accorded to psychons, as is indicated in Figure 6.1. Figure 6.1 has often been wrongly interpreted

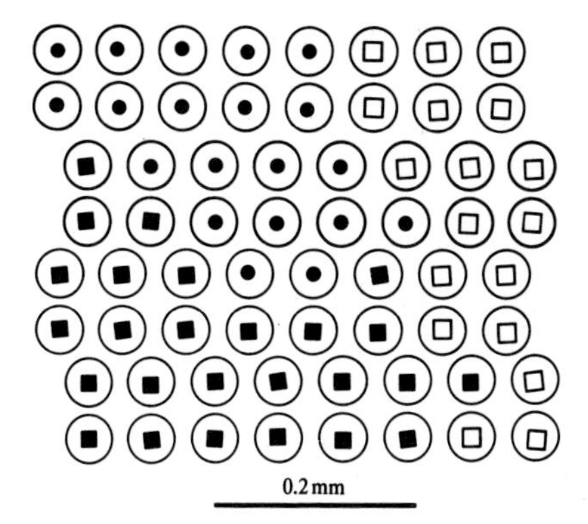

Figure 6.11 A drawing of the postulated patterned arrangement of dendrons as seen from the cortical surface. They have average diameter and spacing (cf. Figure 6.10). Central on each dendron is the symbol of its psychon: *solid squares*, *open squares*, and *circles*. Note the scale of 2 mm.

because World 2 is drawn above the World 1 of the brain. This was for diagrammatic convenience. The reciprocal arrows across the interface show that all the World 2 action is in the neocortex.

This proposed unitary linkage between psychons and dendrons (Figure 6.10) leads to many theoretical developments that in turn will lead to the development of experimental testing procedures. Already there is a vast literature in experimental psychology, cognitive psychology, and neurophysiology that can be assimilated to this unitary theory of mind–brain interaction. For this chapter it is of immediate interest to develop the unitary theory in attempting to explain mind–brain interaction in perception by utilizing quantum mechanics, as has already been done for mental intentions activating dendrons in the SMA (Eccles 1986) (Figure 5.2).

The unitary hypothesis transforms the manner of operation of the intention. If, for example, the psychon for the mental intention is represented by the pattern of open squares on the central dendron of Figure 6.10, it can be seen that the intention acts on the whole dendron with its assembled pyramidal dendrites and their synapses, which could number up to 100 000. So the mental intention would have a large global operation on that dendron. On the unitary hypothesis the psychon would of course operate at each presynaptic vesicular grid (PVG) (see Chapter 6.7) of its dendron in selecting, by means of the quantal probability field, a vesicle for exocytosis. However, collectively there could be tens of thousands of such PVG sites on one dendron, so great amplification is ensured by the unitary operation of the linked psychons and dendrons. One has to recognize that in a lifetime of learning the intention to carry out a particular movement would be channelled largely to those particular psychons that are linked to those dendrons of the neocortex (the SMA) that are appropriate for bringing about the required action.

6.7 The Action of Mental Events on Dendrons

Figure 5.4 shows that the mental events of thinking (psychons) can very effectively activate the neocortex even when no bodily movement is induced. To analyse this action it is essential to define the mode of action of a dendron (Figures 6.8 and 8.2) with large ensembles of apical dendrites of pyramidal cells, each with about one spine synapse per micrometre (Figure 6.4b), so that there can be 100 000 spine synapses on a dendron. The apical dendrites of a dendron are usually closely associated, but not touching (Figure 6.4b), although their spines (Figure 6.4a) may interlace (Feldman 1984).

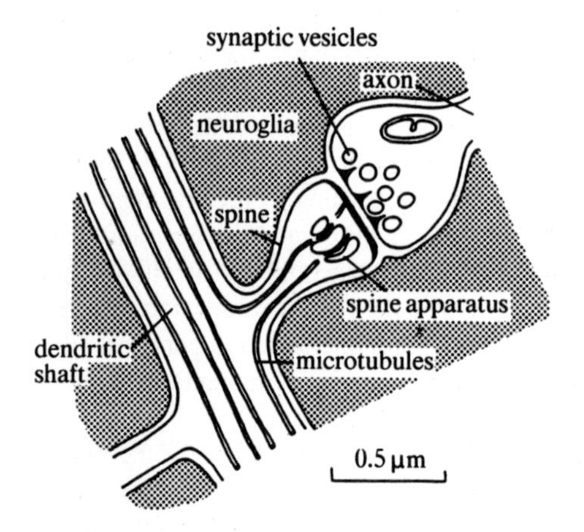

Figure 6.12 A drawing of a synapse on a dendritic spine. The bouton contains synaptic vesicles and dense projections on the presynaptic membrane (Gray 1982).

Figure 6.12 gives a diagram of a spine synapse, showing the nerve fibre expanded to a terminal bouton that makes close contact with a special membrane-thickening of the spine. In the bouton are numerous vesicles, each of which contains 500 to 10 000 molecules of the specific synaptic transmitter substance, which is glutamate or aspartate for the great majority of excitatory boutons in the cerebral cortex. Some synaptic vesicles are in close contact with the presynaptic membrane confronting the postsynaptic membrane across the extremely narrow synaptic cleft. These synaptic vesicles appear to be arranged between dense projections.

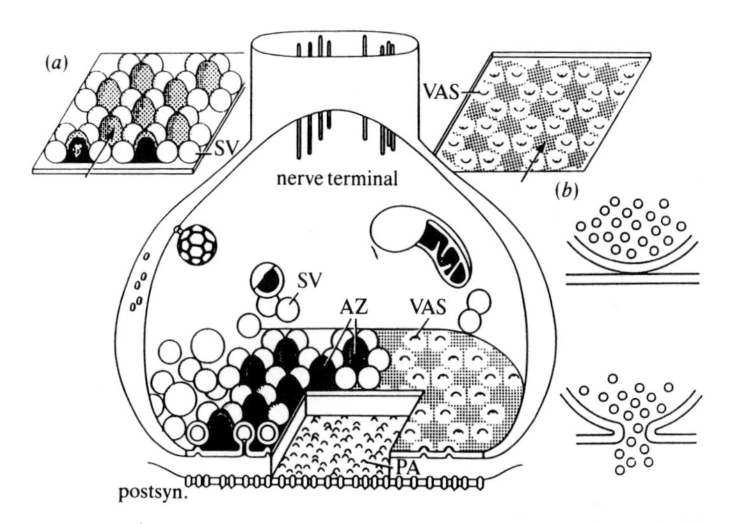

Figure 6.13 (**a**) The schema of the mammalian central synapse. The active zone (AZ) is formed by presynaptic dense projections spacing synaptic vesicles (SV). PA = particle aggregations of postsynaptic membrane (postsyn.). Note the synaptic vesicles in hexagonal array, as is well seen in the *upper-left inset* and the vesicle attachment sites (VAS) in the *right-hand inset*. See the text for further description (see also Akert et al. (1975)). (**b**) Stages of exocytosis with release of transmitter into the synapic cleft (Kelly et al. 1979).

Further structural analysis, particularly by the freeze-fracture technique of Akert et al. (1975), has led to the construction of a diagram of an idealized spine synapse (Figure 6.13a), which is shown in perspective with partial excisions to reveal the deeper structures. The relatively loose arrangement of synaptic vesicles and presynaptic dense projections (Figure 6.12) is replaced in Figure 6.13a as the precise packing illustrated in the inset on the left, with the synaptic vesicles in hexagonal array packaged between the presynaptic vesicular grid (PVG), and it can be regarded as having paracrystalline properties. The boutons of brain synapses have usually a single PVG, as indicated in Figures 6.12 and 6.13a.

There are only approximate figures for the number of synaptic vesicles incorporated in a PVG. The usual number appears to be 30–50. Thus only a very small proportion of the synaptic vesicles of a bouton (about 2 000) are embedded in the firing zone of the PVG. The remainder are loosely arranged in the interior of the bouton, as is partly illustrated in Figures 6.12 and 6.13a.

Figure 6.13b shows at high magnification a part of a synaptic vesicle with its contained transmitter molecules in contact with the presynaptic membrane, as is also seen for two vesicles to the left of Figure 6.13a. Below is the process of exocytosis with release of the transmitter molecules into the synaptic cleft, as is also seen for one vesicle in Figure 6.13a. To the right of Figure 6.13a, after the vesicles and the dense projections have been stripped off, the vesicle attachment sites (VAS) are seen in hexagonal array, as is also seen in the inset diagram to the left.

When a nerve impulse invades a bouton (Figures 6.12 and 6.13), the depolarization causes the entry of Ca^{2+} ions that combine with calmodulin and that may act on an apposed vesicle to trigger an exocytosis. As indicated in Figure 6.13a, there may be 30–50 vesicles incorporated in the PVG, yet in response to a triggering impulse only one sometimes suffers exocytosis (Jack et al. 1981; Korn and Faber 1987). Evidently the exocytosis is subject to control by some unknown holistic property of the paracrystalline PVG.

6.8 A New Hypothesis of Mind–Brain Interaction Based on Quantum Physics: The Microsite Hypothesis

The materialist critics argue that insuperable difficulties are encountered by the hypothesis that immaterial mental events such as thinking can act in any way on material structures such as neurons of the cerebral cortex, as is depicted in Figure 6.1. Such a presumed action is alleged to be incompatible with the conservation laws of physics, in particular with the first law of thermodynamics. This objection would certainly be sustained by the nineteenth-century physicists and by neuroscientists and philosophers who are still ideologically in the physics of the nineteenth century, not recognizing the revolution wrought by quantum physicists in the twentieth century.

In formulating more precisely the dualist hypothesis of mind–brain interaction, the initial statement is that the whole world of mental events (World 2) has an existence as autonomous as the world of matter–energy (World 1) (Figure 6.1). The present hypothesis does not relate to these ontological problems, but merely to the mode of action of psychons on dendrons, that is, to the nature of the downward arrows across the interface in Figure 6.1.

As stated in an earlier publication (Eccles 1986), it is possible to resolve this impasse because the structures concerned in synaptic transmission are so extremely small that they can be operated analogously to the probability fields of quantum physics as described by Margenau (1984). The essential movement in exocytosis is to open a channel, as illustrated in Figure 6.13b. It can be calculated that this involves the displacement of a particle of about 10^{-18} g and not the much larger synaptic vesicle of about 3×10^{-17} g, as was originally proposed (Eccles 1986). Moreover, the vesicles are already in position in the presynaptic vesicular grid (Figure 6.13a, b) so the exocytosis is not dependent on movement through a viscous medium. The postulated action of a psychon would do no more than select for exocytosis any vesicle already in apposition in the paracrystalline PVG with its holistic control as described above.

It can be concluded that calculation on the basis of the Heisenberg uncertainty principle shows that a vesicle of the presynaptic vesicular grid (Figure 6.13a) could conceivably be selected for exocytosis by a psychon acting analogously to a quantal probability field. As indicated in Figure 6.13b the energy required to initiate the exocytosis by a particle displacement could be paid back at the same time and place by the escaping transmitter molecules from a high to a low concentration. In quantum physics at microsites energy can be borrowed provided it is paid back at once. So the transaction of exocytosis need involve no violation of the conservation laws of physics.

Before these latest dendron–psychon developments, there was the problem of the order of magnitude of a mental intention that was acting by selection for exocytosis of a single vesicle. A great amplification was essential. However, on the dendron–psychon hypothesis a psychon has as its field for exocytotic selection the 100 000 spine synapses on its dendron (Figure 6.10), which gives a possible amplification of its action by several orders of magnitude. Furthermore, it can be assumed from the enormous numbers of activated dendrons in Figure 6.11 that many adjacent dendrons have closely related psychons, as has been shown in Figure 6.11, where there are assemblages of three distinct varieties of dendron–psychon to form three modules for transmission.

6.9 How Neuronal Activity in the Sensory Systems Could Evoke Conscious Perceptions

Hitherto, for the whole of perceptual experience, there has been no microsite hypothesis. The additional hypothesis for all the upward arrows of Figure 6.1 can be developed in stages. Perception is dependent on a *directed attention.* As seen above in Figure 5.4a, a mental attention to some surface of the body activates the neocortical areas specifically related to that area and also more widely to the frontal lobe. No special hypothesis is needed beyond the microsite hypothesis already developed for the action of intention on the dendrons of the SMA (Figure 5.2b).

The response of the neocortex to attention (Figure 5.4) is preparatory to the transaction whereby dendrons are activated in the perceptual process to produce the perceptual mental events. For example, it can be asked: how can activated dendrons of the tactual system give rise to some specific tactile perception? This is the problem of the reverse arrows from World 1 to World 2 of Figure 6.1.

Let us concentrate on the attentional act whereby psychons are exciting dendrons (Figure 5.4a) in accordance with the microsite hypothesis. On to that background there is superimposed an activation of the dendrons by some perceptual input, for example a tactile input, which could specifically excite the apical dendrites of the dendron linked to the right psychon of Figure 6.10 (solid circles) that gives a tactual perception. So that psychon is presented with an increase in its dendron of vesicles available for exocytosis in accord with selection by means of the quantal probability field. The hypothesis is that each such exocytosis is a 'success' for the psychon, which gives a signal that is transmitted into the mental world, World 2 of Figure 6.1.

The sequence for tactual perception would be hypothesized as follows.

1 Background activation by attention to the tactual area (Figure 5.4).
2 Sensory input into the tactual nervous system.
3 Activation in the neocortex of the dendrons of the tactual system.
4 Increased exocytosis from the presynaptic vesicular grids of the pyramidal cells of these dendrons. This gives increased opportunity for selective exocytosis by the linked psychon (cf. Figures 6.10 and 6.11, solid circles), which is in accord with a quantum probability field.
5 The increase in vesicular selection by the psychon for touch gives directly the experience of a tactual perception in World 2 and a psychon 'success' signal for transmission and integration in World 2.

All other perceptions of the outer sense in Figure 6.1 can be similarly explained by appropriate attention.

This unitary perceptual hypothesis is inadequate because it is limited to specifically linked neural–mental units (Figure 6.10). There is no explanation of the tremendous enigma of the unification of our perceptual experiences. For example, from some dynamic activity pattern of millions of visual psychons we perceive a visual picture with all its qualities and movement. It could be that an explanation may emerge from the psychon integration of the diverse activities of dendrons in stage after stage of the visual processing system.

It would have to be assumed that at each stage the psychons are dependent on the dendrons. Possibly diagrams such as those of Figures 6.10 and 6.11 are applicable even to the highest levels of the neocortex with its gnostic functions (Eccles 1989, Chapter 9). Alternatively, some psychons may be linked only with psychons (Chapter 10 of the present book).

6.10 The Mental World of Psychons (World 2)

An extension of the microsite hypothesis of mind–brain interaction (Eccles 1986) has led to some extraordinary developments, which are as yet very tentative. The original microsite hypothesis used quantum physics in explaining how a non-material mental event, an intention to move, can cause microsite activity across the interface between mind and brain, largely in the SMA (Roland et al. 1980) (Figure 5.2b). The attempt to develop this hypothesis for the brain–mind problem in perception has necessitated a radically new hypothesis.

In the original microsite hypothesis (Chapter 5) the mental intention acted in accordance with quantum physics (Margenau 1984) to select for exocytosis (Figure 4.6b) a vesicle of the activated presynaptic vesicular grid (Figure 5.2b). It was a unitary action at a microsite and had to be enormously amplified by assuming that there were thousands of microsites on that dendrite and on dendrites of many adjacent pyramidal cells.

In this present unitary hypothesis the linked dendrons and psychons are central to the act. Thus the mental intention acting through a psychon (cf. Figures 6.10 and 6.11) automatically has available tens of thousands of activated PVGs with their vesicles awaiting selection.

In the reverse transaction (perception), brain to mind, it is necessary to have an extension of the hypothesis, namely that every time a psychon successfully selects a vesicle for exocytosis (in accord with the quantal probability field) the 'micro-success' is registered in the psychon for transmission through the mental world (World 2 of Figure 6.1). There would, of course, be great amplification when the psychon successfully selected, at about that time, large numbers of vesicles from the tens of thousands of PVGs of its dendron. The 'success' signal of the psychon would carry into World 2 the special experiential character of that psychon for integration into the psychon world.

A tentative explanation may be offered for the observation that an input into the sensory nervous system can give rise to a sensory experience. Activation of an appropriate dendron of area V4 (cf. Zeki 1973) can be exemplified by the dendron to the left of Figure 6.10. This can result in a 'success' response of a psychon indicated by the pattern solid squares and so to the experience of a red colour. However, for such an experience it is likely that there are 'success' responses in several adjacent psychons, as is indicated by the ensemble in Figure 6.11.

6.11 General Considerations

It should be recognized that there is superb design of the dendrons for their receptive function, both neuronal and psychic. It must be accepted that all mammals are conscious beings with some control of their actions and some conscious experiences (Eccles 1989, p. 173, and Chapter 10.7 of the present book). The dendron–psychon interaction is thus essential to their mental life. The human situation is a further development with the coming of self-consciousness (Eccles 1989, pp. 203 and 218), in which psychons may exist apart from dendrons in a unique psychon world, which is the world of the self (Figure 6.1). There are great unknowns in this postulated world of psychons. Their very nature is to give experiences, and we can only indicate their existence diagrammaticaly by the site of their dendron action, which is shown as an ensheathing in Figure 6.10.

Transmission of psychon to psychon could explain the unity of our perceptions and of the inner world of our mind that we continually experience from moment to moment: that is, for all of the World 2 experiences illustrated in Figure 6.1, above the interface. This problem has been considered in Chapter 6.10 in attempting to explain the unity of visual experience. Hitherto it has been beyond explanation by any mind–brain theory that multifarious neural events in our cerebral cortex can from moment to moment give us global mental experiences that have a unitary character. We feel central to our experiential world (World 2). This phenomenon is shown in the central core of World 2 in Figure 6.1 with its labelling of psyche, self, soul. Arrows are shown projecting into this central locus from the region of outer sense and inner sense. This raises a fundamental question: are the experiences of the self also composed of unitary psychons in the same manner as for perceptual experiences, for example? If so, is each of these psychons also linked with its dendron, and where in the neocortex are these dendrons? We can further ask if there is a category of organized psychons not linked to dendrons, but only with other psychons, forming a psychic entity apart from the brain, as will be discussed in Chapter 10.7.

References to Chapter 6

Akert, K., Peper, K., and Sandri, C. (1975) Structural organization of motor end plate and central synapses, in *Cholinergic Mechanisms*, edited by P. G. Waser (Raven, New York), pp. 43–57.

Armstrong, D. M. (1981) *The Nature of Mind* (Cornell University Press, Ithaca NY).

Bunge, M. (1980) *The Mind–Brain Problem* (Pergamon, Oxford).

Dennett, D. C. (1969) *Content and Consciousness* (Routledge and Kegan Paul).

Dennett, D. C. (1978) *Brainstorms* (Bradford Books/MIT Press, Cambridge MA).

Eccles, J. C. (1986) Do mental events cause neural events analogously to the probability fields of quantum mechanics?, *Proc. Roy. Soc. London* B **227**, 411–428.

Eccles, J. C. (1989) *Evolution of the Brain: Creation of the Self* (Routledge, London).

Feigl, H. (1967) *The 'Mental' and the 'Physical'* (University of Minnesota Press, Minneapolis).

Feldman, M. (1984) Morphology of the neocortical pyramidal neuron, in *Cerebral Cortex*. Vol. 1. *Cellular Components of the Cerebral Cortex*, edited by A. Peters and E. G. Jones (Plenum, New York), pp. 123–200.

Feldman, M., and Peters, A. (1974) A study of barrels and pyramidal dendritic clusters in the cerebral cortex, *Brain Res.* **77**, 55–76.

Fleischhauer, K., Petsche, H., and Wittkowski, W. (1972) Vertical bundles of dendrites in the neocortex, *Z. Anat. Entwickl. Gesch.* **136**, 213–223.

Goldman, P. S., and Nauta, W. J. H. (1977) Columnar distribution of corticocortical fibers in the frontal association, limbic and motor cortex of the developing rhesus monkey, *Brain Res.* **122**, 393–413.

Gray, E. G. (1982) Rehabilitating the dendritic spine, *Trends Neurosci.* **5**, 5–6.

Hamlyn, L. H. (1963) An electron microscope study of pyramidal neurons in the Ammon's Horn of the rabbit, *J. Anat.* **97**, 189–201.

Hebb, D. O. (1980) *An Essay on Mind* (Lawrence Erlbaum, Hillsdale).

Jack, J. J. B., Redman, S. J., and Wong, K. (1981) The components of synaptic potentials evoked in cat spinal motoneurones by impulses in single group Ia afferents, *J. Physiol., London* **321**, 65–96.

Kelly, R. B., Deutsch, J. W., Carlson, S. S., and Wagner, J. A. (1979) Biochemistry of neurotransmitter release, *A. Rev. Neurosci.*, **2**, 399–446.

Korn, H., and Faber, D. S. (1987) Regulation and significance of probabilistic release mechanisms and central synapses, in *New Insights into Synaptic Function*, edited by G. M. Edelman, W. E. Gall, and W. M. Cowan (Neurosciences Research Foundation) (Wiley, New York), pp. 57–108.

Margenau, H. (1984) *The Miracle of Existence* (Ox Bow, Woodbridge CT).

Peters, A., and Kara, D. A. (1987) The neuronal composition of area 17 of rat visual cortex. IV. The organization of pyramidal cells, *J. Comp. Neurol.* **260**, 573–590.

Popper, K. R., and Eccles, J. C., (1977) *The Self and Its Brain* (Springer, Berlin, Heidelberg).

Ramón y Cajal, S. R. (1911) *Histologie du Système Nerveux de l'Homme et des Vertébrés*. Vol. 2 (Maloine, Paris).

Roland, P. E. (1981) Somatotopical tuning of postcentral gyrus during focal attention in man. A regional cerebral blood flow study, *J. Neurophysiol.* **46**, 744–754.

Roland, P. E., Larsen, B., Lassen, N. A., and Skinhøj, E. (1980) Supplementary motor area and other cortical areas in organization of voluntary movements in man. *J. Neurophysiol.* **43**, 118–136.

Roland, P. E., and Friberg, L. (1985) Localization of cortical areas activated by thinking, *J. Neurophysiol.* **53**, 1219–1243.

Ryle, G. (1949) *The Concept of Mind* (Hutchinson, London).

Schmolke, C. (1987) Morphological organization of the neuropil in laminae II–V of rabbit visual cortex, *Anat. Embryol.* **176**, 203–212.

Schmolke, C., and Fleishhauer, K. (1984) Morphological characteristics of neocortical laminae when studied in tangential semithin sections through the visual cortex of the rabbit, *Anat. Embryol.* **169**, 125–132.

Searle, J. (1984) *Minds, Brains and Science* (British Broadcasting Corporation, London).

Szentágothai, J. (1975) The 'module concept' in cerebral cortex architecture, *Brain Res.* **95**, 475–496.

Szentágothai, J. (1978) The neuron network of the cerebral cortex. A functional interpretation, *Proc. Roy. Soc. London* B **201**, 219–248.

Szentágothai, J. (1979) Local neuron circuits of the neocortex, in *The Neurosciences Fourth Study Program*, edited by F. O. Schmitt and F. G. Worden (MIT Press, Cambridge MA), pp. 399–415.

Valverde, F. (1968) Structural changes in the area striate of the mouse after enucleation, *Expl. Brain Res.* **5**, 274–292.

Zeki, S. M. (1973) Colour coding in rhesus monkey prestriate cortex, *Brain Res.* **53**, 422–427.

7 The Evolution of Consciousness

7.1 Introduction

For consciousness to be experienced in a hitherto mindless world there had to be a neocortex that was greatly evolved, as it came to be in the mammalian brain, even of the primitive insectivores (Figure 7.3). The contrast with the reptilian brain (Figure 7.4) and the avian brain (Figure 7.5) revealed the mammalian evolution to a neocortex and so to consciousness, which is the most wonderful happening of the whole evolutionary story. This is only slowly being appreciated after the long dark night of behaviourism and materialism, as described in this chapter, and also by Searle (1992) (see Chapter 3.9).

In the past decade or so there has been a general recognition of the centrality of consciousness in human experiences (Armstrong 1981; Dennett 1978; Searle 1984; Ingvar 1990; Edelman 1989; Penrose 1989; Crick and Koch 1990; Squires 1988; Hodgson 1991). The mental word 'consciousness' is now an 'in' word, being used shamelessly even by strong materialists!

An introductory statement for the evolution of consciousness is that one cannot expect that consciousness came to higher animals as a sudden illumination. Rather, as with life originating in a prebiotic world, it would be anticipated that consciousness came secretly and surreptitiously into a hitherto mindless world. Moreover, as we attempt to discover evidence for consciousness from the study of animal brains and behaviour, we can only assess probability. We search for manifestations of consciousness in mammals because they appear to have similar experiences, such as feelings and pain. We recognize that consciousness is central to the qualia that fill our waking life like a rich tapestry replete with feelings, thoughts, memories, imaginings, and sufferings. Our experience is uniquely ours, but we are rescued from solipsism by communication with other human beings by language and other subtle creations, such as music and gesture, and by sharing our immersion in a rich inherited culture.

7.2 The Mammalian Cerebral Cortex

Mammals have a cerebral cortex qualitatively similar to ours, though, with rare exceptions, much smaller. Some exhibit intelligence and a learned behaviour and are moved by feelings and moods, even with emotional attachment and understanding. So we must give them some feelings and primitive qualia such as we human beings experience even though it cannot be rationally established in the way that is possible by interhuman communication (Eccles 1990a).

I am presenting a biological basis for an evolutionary origin of consciousness. It derives from a hypothesis of mind–brain interaction that has already been published (Eccles 1990a, b and Chapters 5 and 6) and that is based on the special anatomical and functional properties of the mammalian cerebral cortex. The microproperties of neural communication in the cerebral cortex (Figure 7.1a) are in classical physics and are not of immediate concern in mind–brain interaction. Rather, our concern is in the ultramicroproperties (Figure 7.2b), where quantum physics may be expected to play a key role (Eccles 1986, 1990b; Stapp 1992; Beck and Eccles 1992 and Chapter 9).

A pyramidal cell of the mammalian cerebral cortex (Figure 7.1a) has on its apical dendrite thousands of excitatory spine synapses (Figures 7.1b and 7.2a). Each of these synapses (Figure 7.2b) operates through a presynaptic vesicular grid (Akert, Peper, and Sandri 1975) of 30–50 synaptic vesicles filled with molecules of the synaptic transmitter substance (Figure 7.2b). Each vesicle is poised on the presynaptic membrane (Akert, Peper, and Sandri 1975) for emission of its transmitter molecules in exocytosis (Figure 7.2b–d).

Exocytosis is the basic unitary activity of the cerebral cortex. Each all-or-nothing exocytosis results in a brief excitatory postsynaptic depolarization, the excitatory postsynaptic potential (EPSP). Summation (Figure 4.1) of many hundreds of these milli-EPSPs is required for an EPSP large enough to generate the discharge of an impulse by the pyramidal cell. This impulse will travel along its axon (Figure 7.2a) to make effective excitation at its many synapses (Figure 7.1a). This is the conventional macrooperation of the neural component of the neocortex and it can be satisfactorily described by classical physics even in the most complex designs of network theory (Mountcastle 1978) and neuronal group selection (Edelman 1989).

A quite different theory is necessary for describing the manner in which mental events could generate neural events of the neocortex. Mental activity of the neocortex results in an increased metabolism that causes local increases in cerebral blood flow, as shown by radio-xenon mapping by

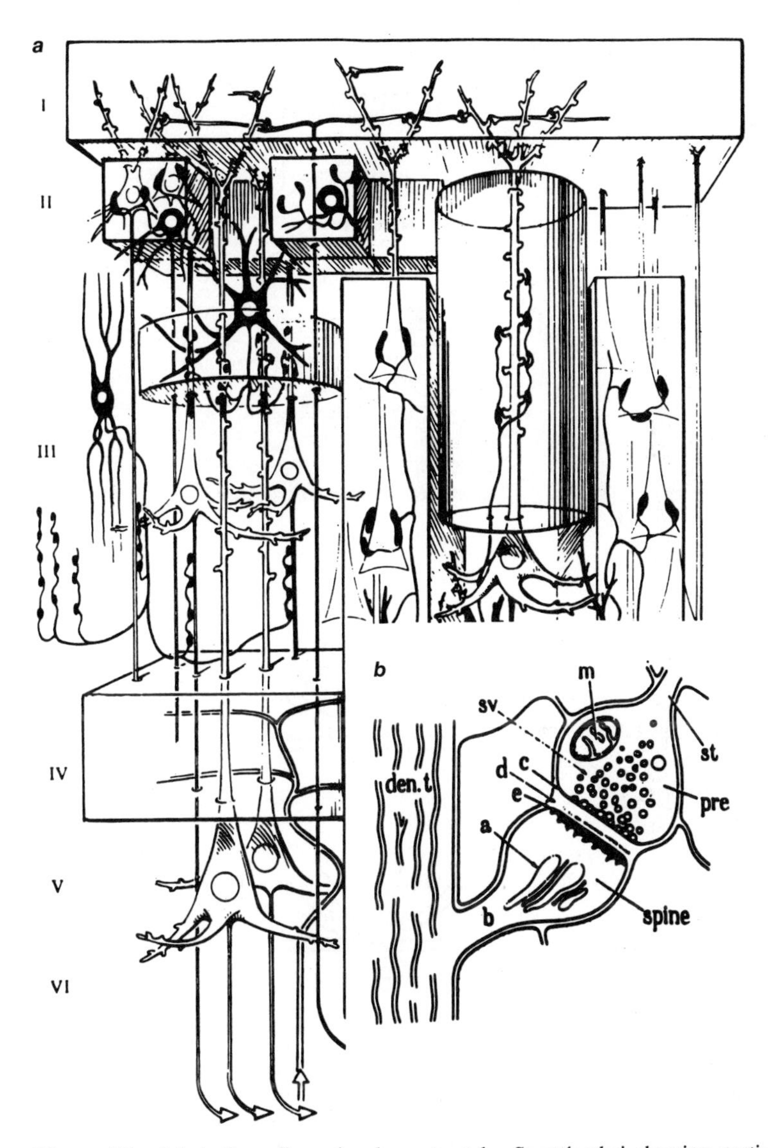

Figure 7.1 (**a**) A three-dimensional construct by Szentágothai showing cortical neurons of various types. There are two pyramidal cells in lamina V and three in lamina III, with one shown in detail on the right (Szentágothai 1978). (**b**) The detailed structure of a spine synapse on a dendrite (den). st = axon terminating in a synaptic bouton or presynaptic terminal (pre); sv = synaptic vesicles; c = presynaptic vesicular grid; d = synaptic cleft; e = postsynaptic membrane; a = spine apparatus; b = spine stalk; m = mitochondrion (Gray 1982).

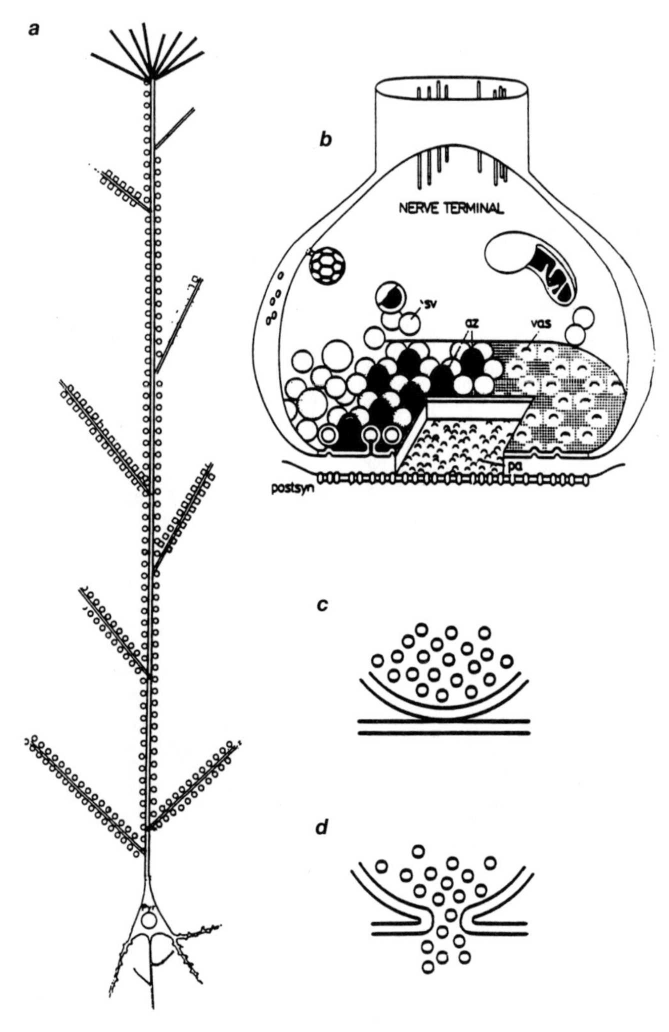

Figure 7.2 (**a**) A drawing of a lamina V pyramidal cell (Pyr), with its apical dendrite showing the side branches and the terminal tuft all studded with spine synapses and boutons (not all shown). The soma with its basal dendrites has an axon with axon collaterals before leaving the cortex. (**b**) The schema of a mammalian central synapse. The active zone (az) is formed by presynaptic dense projections spacing synaptic vesicles (sv). pa = particle aggregations of postsynaptic membrane (postsyn). Note the synaptic vesicles in hexagonal array and the vesicle attachment sites (vas) on the right (Akert, Peper, and Sandri 1975). (**c, d**) Stages of a synaptic exocytosis: the vesicle is shown with the release of transmitter molecules into the synaptic cleft (Kelly, Deutsch, Carlson, and Wagner 1979).

Ingvar (1990) and Roland et al. (Roland, Larsen, Lassen, Skinhøj 1980) (Figure 5.2), or by the more accurate positron-emission tomography (PET) scanning technique of Raichle and associates (Posner, Petersen, Fox, and Raichle 1988)(Chapter 10.5) for the activity of the human brain in response, for example, to the mental demand to generate words (Figure 10.3d).

These mental actions can be attributed to an increase in exocytoses. A presynaptic impulse propagating into a bouton (Figure 8.5) and causing an influx of Ca^{2+} ions generates an exocytosis not on every occasion but with a wide range of probability, often 1 in 3 or 1 in 4, and never more than one exocytosis (Jack, Redman, and Wong 1981; Korn and Faber 1990; Redman 1990; Sayer et al. 1989, 1990). This controlling effect probably is due to the vesicles being embedded in the paracrystalline presynaptic vesicular grid (Figure 7.2b) and thus being subject to quantum physics (Beck and Eccles 1992 and Chapter 9).

7.3 Dendrons and Psychons

The apical dendrites of the pyramidal cells in laminae V, III, and II bundle together as they ascend to lamina I (Schmolke and Fleischhauer 1984; Peters and Kara 1987) (Figures 6.8 and 8.2), so neural receptor units of the cerebral cortex are formed, being composed of approximately 100 apical dendrites plus their branches (Figure 8.2b). They are called dendrons, three of which are drawn in Figure 6.10 for apical dendrites of lamina V pyramidal cells.

In the hypothesis of mind–brain interaction (Eccles 1986, 1990a, b; Beck and Eccles 1992; and Chapters 6 and 9) it is proposed that the whole mental world is microgranular, with the mental units being called psychons. Ideally, there would be one psychon for each dendron, as shown in Figure 6.10 by the ensheathing of each of the three dendrons by the superposed patterns of solid squares, open squares, and solid circles. It is further proposed that mind–brain interaction occurs for each psychon–dendron unit and that it can be accounted for by quantum physics (Eccles 1990a, b; Beck and Eccles 1992 and Chapter 9).

There is an immense input to a dendron by thousands of synapses on the apical dendrites plus side branches of each pyramidal cell (Figure 7.2a), which gives tens of thousands of synaptic vesicles on the presynaptic vesicular grids of one pyramidal cell (Figure 7.2b) that are poised for exocytosis. Thus there are millions of poised synaptic vesicles on a dendron (Figure 8.2b), which consequently is a tremendously sensitive receptor for psychon inputs. And so consciousness could be experienced (Beck and Eccles 1992 and Chapter 9).

7.4 Cerebral Evolution and Consciousness: The Hypothesis

In all mammals so far examined, there is the same composition of apical dendrites of pyramidal cells arranged in dendrons (Figures 6.10 and 8.2b). There may be as many as 40 million dendrons for the human cerebral cortex (Eccles 1990b and Chapters 5 and 6), but there are probably no more than 200 000 for the cerebral cortices of the most primitive mammals, the basal insectivores, that may have dendron assemblages characteristic of higher mammals. This estimate is based on measurements by Stephan et al. (1991) of the neocortices of more than 50 species of insectivores, including 4 basal insectivores. As illustrated in Figure 7.3, in the neocortex the pyramidal cells appear to have a vertical orientation with a dendron structure (Schmolke 1993).

The cerebral cortex with the synaptic machinery of its dendrons can be regarded as a functionally effective neural design that evolved in natural selection as a purely material structure for efficient performance of the cerebral cortex in integrating the increased complexity of neural inputs that resulted from the evolutionary advances in receptors for special senses—light, sound, touch, movement, and olfaction. The hypothesis is that biological evolution induced in the neocortex the design of apical dendrites that is recognized as a dendron (Figures 6.8, 6.10, and 8.2) and that had as a side effect the capacity for amplifying the minute psychon effects induced by the mind. And so psychons came to exist. Thus it is sufficient for the emergence of consciousness. It will be noted that the hypothesis is restricted to the role of the cerebral cortex in providing an explanation of how in evolution the cerebral cortex came to interact with the 'mind world'. The actual experiences, qualia, provided by the mind-world psychons—light, colour, sound, touch, taste, smell, pain, intentions, feelings, and memories—in all of their uniqueness are not explained. The most we can say is that the dendron–psychon linkage is related to the types of qualia experienced, since the hypothesis is that each psychon is a unitary experience in consciousness. We may ask if the mind world existed before it could be experienced by the evolving cerebral cortices of primitive mammals. The answer should be that the mind world came to exist when the evolving cerebral cortex had the microsites with synaptic vesicles poised in the presynaptic vesicular grids (Figures 7.1b and 7.2b). This microstructure emerged with the evolving dendrons that could interact with psychons as described above.

How far down our evolutionary origin can we recognize some evidence for consciousness? If all mammals experience consciousness, its evolution-

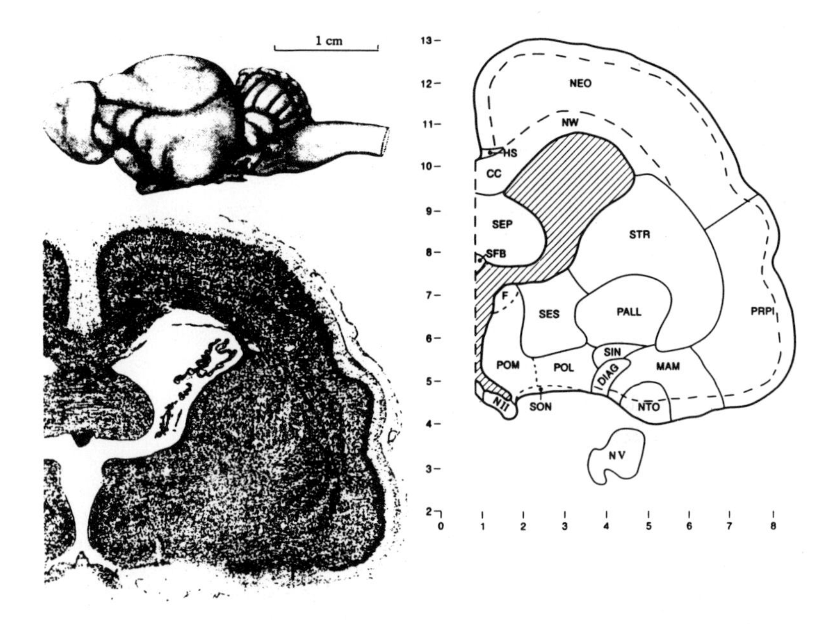

Figure 7.3 (**a**) The brain of an insectivore neocortex above the prepyriform cortex. (**b**) A transverse section of the brain of an insectivore (a hedgehog). (**c**) A diagram of the transverse section shown in (b) showing the following: CC = corpus callosum; DIAG = diagonal band of Broca; F = fimbria–fornix complex; HS = hippocampal supracommissuralis; MAM = medial amygdaloid division; NEO = neocortex; NTO = nucleus of the preoptic area; POM = medial preoptic area; PRPI = prepyriform region; SEP = telencephalic septum; SES = stria terminalis nuclei; SFB = subfornical body; SON = suprapotic nucleus; STR = striatum. *Scale numbers* are mm (Stephan, Baron, and Frahm 1991).

ary origin can be as early as 200 million years before the present (Jerison 1990).

Now comes the question about the reptilian origin of mammals (Jerison 1990; Ulinski 1990). The forebrain of reptiles (Figure 7.4) shows an undeveloped cerebral cortex, which is in great contrast to the most primitive mammalian cortex (Figure 7.3). Clearly, the reptilian cerebral cortex has far to evolve before it can play a significant role in cerebral activity and in mind–brain interaction based on dendrons. But we have to recognize that the primitive reptilian cerebral cortex (Figure 7.4) did, eventually, evolve into a primitive mammalian cortex (Figure 7.3).

There seems to be no good evidence that reptiles experience consciousness. Thorpe (1974) reports on the behaviour of amphibia and reptiles, but

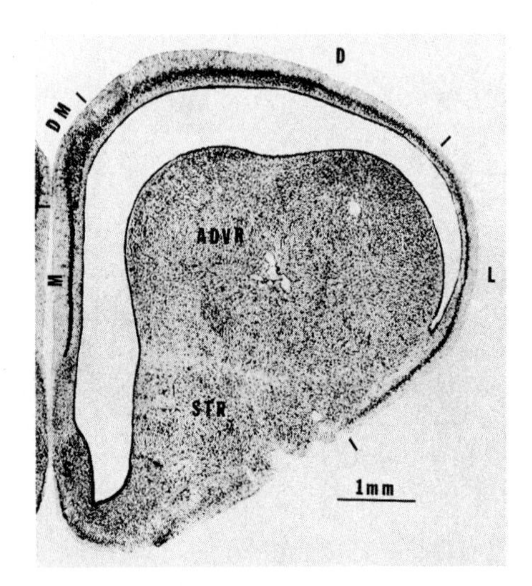

Figure 7.4 A Nissl section from the encephalon of the alligator *Alligator mississippiensis*. The cortex is divided into the medial cortex (M), dorsomedial cortex (DM), dorsal cortex (D), and lateral cortex (L). ADVR = anterior dorsal ventricular ridge; S = septum; STR = striatum (Ulinski 1990).

does not mention consciousness. All actions can be regarded as instinctive and learned, and this has to be assumed also for still lower vertebrates—the fish.

So consciousness appears to have come into the mindless world of biological evolution with the origin of mammals. Animals would not experience 'gleams' of consciousness (feelings) until the mammalian cerebral cortex with its microsites of neuronal structures (dendrons) had evolved with the propensity for relating to a world other than the matter–energy world. So psychons, with their conscious experiences, entered the hitherto mindless world.

The hypothesis that consciousness gives an animal behaviour (Hodgson 1991) that is of great significance in natural selection could be tested by comparative studies on reptiles and mammals in the same environment.

The reptilian–avian transition is difficult to follow (Ulinski and Margoliash 1990). The medial, dorsomedial, and lateral components of the reptilian cortex (Figure 7.4) continued in a relatively reduced form in birds. However, birds have a unique development in the forebrain, called the Wulst (Figure 7.5), that is concerned with vision. I would follow Thorpe (1978) in agreeing that many birds exhibit 'insightful' behaviour particularly at feeding time, and that is evidence of conscious experiences. So there is

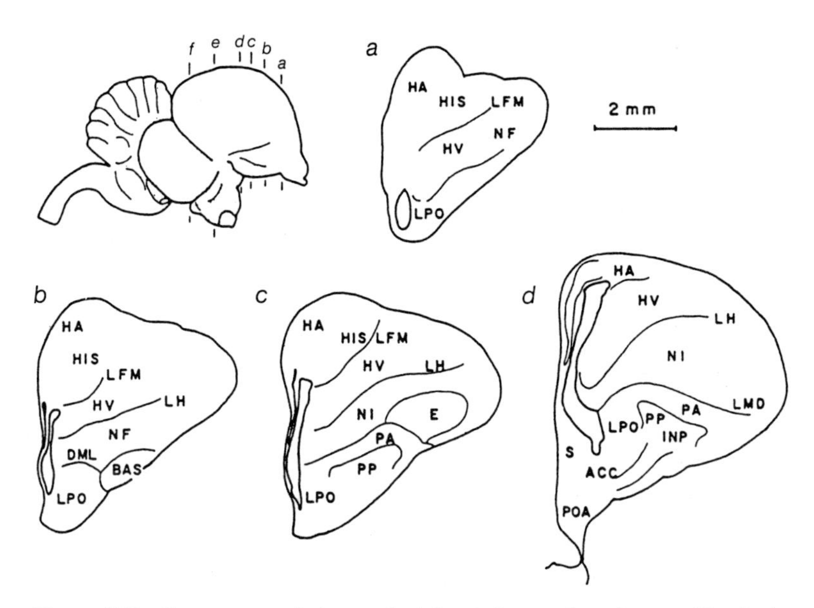

Figure 7.5 Components of the cerebral hemispheres of a pigeon. The basic anatomy of the telencephalon of birds is illustrated by a series of transverse sections through the right cerebral hemisphere. The planes of the sections are illustrated in the inset (*upper left*), which shows a lateral view of a pigeon brain. ACC = nucleus accumbens; BAS = nucleus basalis; DML = dorsomedial lateral nuclei; E = ectostriatum; HA = hyperstriatum accessorium; HIS = hyperstriatum intercalatus superioris; HV = hyperstriatum ventrale; INP = intrapeduncular nucleus; L = field L; LFM = frontal medullary lamina; LH = lamina hyperstriatica.

the problem of identifying the ultramicrosites in the bird pallium that could be involved in mind–brain interaction, giving birds gleams of conscious experiences (Figure 7.5).

The special structure (the Wulst) includes the various components (HA, HIS, HV, and LH) of the hyperstriatum shown in Figure 7.5 and is specially related to the visual brain. It is the most rostral telencephalon. Although there is indication that the Wulst is the most advanced component of the avian brain, much more detailed information is required before one can relate the Wulst to the presumed avian consciousness.

Even the most evolved brains of molluscs, i.e. the octopus or aplysia, have far too simple a brain to provide conscious experiences on the level of the insectivore brain or even the avian Wulst.

Hitherto the matter–energy world had been globally sufficient in a mindless universe. Now we have an evolutionary explanation for the phylogenetic origin of mammalian consciousness. It would occur initially in the primitive cerebral cortices of evolving mammals, such as the basal insectivores of today. The evolution of a neurally efficient cerebral cortex had as an 'unintended' consequence a unique property—that of interacting by quantum physics with the mind world (Beck and Eccles 1992 and Chapter 9). It is a superb example of what I have called anticipatory evolution (Eccles 1990a).

7.5 Conclusions

It should be pointed out that the present hypothesis of the evolutionary origin of consciousness has four great attractions: (i) that it is neuroanatomical, (ii) that it is in accord with biological evolution, (iii) that it utilizes the most highly evolved structures of the cerebral cortex with their ultramicrosite operation, and (iv) that it is based on quantum physics.

The quantum physicist Stapp (1993)(see Chapters 3.13 and 3.14) has maintained that classical physics provides an unacceptable basis for the hypotheses of brain–mind interaction, which instead requires quantum physics. So Stapp (1992) rejects such theories as the neuronal network (Mountcastle 1978) and neuronal group selection (Edelman 1989) and proposes a theory based on quantum physics. This theory needs a secure basis in ultramicrosite neuroscience of the cerebral cortex where quantum physics could be effective. This has now been done by Beck and myself (Beck and Eccles 1992 and Chapter 9).

There has recently been much attention given to the neuronal mechanisms that cause interactions between widely separated cortical areas; this has been named the binding problem (Crick and Koch 1990; Damasio 1989; Singer, Artola, Engel, and Kreiter 1991). It is here suggested that global conscious experiences given by psychons in their interactions could play an important role in the binding problem.

As has been described (Eccles 1990a, b and Chapter 5), consciousness gives a global experience (Hodgson 1991) from moment to moment of the diverse complexities of cerebral performances—e.g. it would give a mammal global experiences of a visual world for guiding its behaviour far beyond what is given by the robotic operation of the visual cortical areas per se.

Thus, conscious experiences would give evolutionary advantage. This simple consciousness need not be an enduring entity, but may merely exist from moment to moment according to the activities of the cerebral cortex. The brain would provide an enduring memory that gives the animal a continuity of behaviour and of feeling of existence. Nevertheless, the conscious experiences can be assumed to include quite complicated qualia of devoted attachment and enjoyment as well as pain.

This is a formulation of a Darwinist hypothesis for the origin of consciousness in the most primitive mammals. This hypothesis for the origin of consciousness does not account, however, for the highest levels of consciousness in *Homo sapiens sapiens* (Eccles 1990a, b and Chapter 5; Popper and Eccles 1977)—self-consciousness—which is the unique experience of each human self (Eccles 1990a, b and Chapter 5).

The hypothesis is that in mammalian evolution dendrons evolved for more effective integration of the increased complexity of sensory inputs. These evolved dendrons had the capacity for interacting with psychons that came to exist, so forming the mental world and giving the mammal conscious experiences. In Darwinian evolution, consciousness would have occurred initially some 200 million years ago in relation to the primitive cerebral cortices of evolving mammals. It would give global experiences of a surrounding world for guiding behaviour beyond what is given by the unconscious operation of sensory cortical areas per se. So conscious experiences would give mammals evolutionary advantage over the reptiles, which lack a neocortex giving consciousness. The Wulst of the avian brain needs further investigation to discover how it could give birds the consciousness that they seem to have.

References to Chapter 7

Akert, K., Peper, K., and Sandri, C. (1975) in *Cholinergic Mechanisms*, edited by P. G. Waser (Raven, New York), pp. 43–57.

Armstrong, D. M. (1981) *The Nature of Mind* (Cornell University Press, Ithaca NY).

Beck, F., and Eccles, J. C. (1992) *Nature* (submitted).

Crick, F., and Koch, C. (1990) *Seminars in the Neurosciences* **2**, 263–275.

Damasio, A. R. (1989) *Neural Computat.* **1**, 123–132.

Dennett, D. C. (1978) *Brainstorms* (MIT Press, Cambridge MA), p. 353.

Eccles, J. C. (1986) *Proc. Roy. Soc. London B* **240**, 433–451.

Eccles, J. C. (1990a) in *The Principles of Design and Operation of the Brain*, edited by J. C. Eccles and O. Creutzfeldt (Experimental Brain Research, Series 21) (Springer, Berlin, Heidelberg), pp. 549–568.

Eccles, J. C. (1990b) *Proc. Roy. Soc. London B* **227**, 411–428.

Edelman, G. M. (1989) *The Remembered Present: A Biological Theory of Consciousness* (Basic Books, New York), p. 346.

Gray, E. G. (1982) *Trends Neurosci.* **5**, 5–6.

Hodgson, D. (1991) *The Mind Matters* (Clarendon, Oxford), p. 482.

Ingvar, D. H. (1990) in *The Principles of Design and Operation of the Brain*, edited by J. C. Eccles and O. Creutzfeldt (Experimental Brain Research, Series 21) (Springer, Berlin, Heidelberg), pp. 433–453.
Jack, J. J. B., Redman, S. J., and Wong, K. (1981) *J. Physiol.* **321**, 65–96.
Jerison, H. J. (1990) in *Cerebral Cortex*, Vol. 8A, edited by E. G. Jones and A. Peters (Plenum, New York), pp. 285–309.
Kelly, R. B., Deutsch, J. W., Carlson, S. S., and Wagner, J. A. (1979) *A. Rev. Neurosci.* **2**, 399–446.
Korn, H., and Faber, D. S. (1990) *J. Neurophysiol.* **63**, 198–222.
Mountcastle, V. B. (1978) in *The Mindful Brain*, edited by F. O. Schmitt (MIT Press, Cambridge MA), pp. 7–50.
Penrose, R. (1989) *The Emperor's New Mind* (Oxford University Press, Oxford), p. 466.
Peters, A., and Kara, D. A. (1987) *J. Comp. Neurol.* **260**, 573–590.
Popper, K. R., and Eccles, J. C. (1977) *The Self and Its Brain* (Springer, Berlin, Heidelberg), p. 597.
Posner, M. I., Petersen, S. E., Fox, P. T., and Raichle, M. E. (1988) *Science* **240**, 1627–1631.
Redman, S. J. (1990) *Physiol. Rev.* **70**, 165–198.
Roland, P. E., Larsen, B., Lassen, N. A., and Skinhøj, E. (1980) *J. Neurophysiol.* **43**, 118–136.
Sayer, R. J., Redman, S. J., and Andersen, P. (1989) *J. Neurosci.* **9**, 845–850.
Sayer, R. J., Friedlander, M. J., and Redman, S. J. (1990) *J. Neurosci.* **10**, No. 3, 626–636.
Schmolke, C., and Fleischhauer, K. (1984) *Anat. Embryol.* **169**, 125–132.
Schmolke, C. (1993) Personal communication.
Searle, J. (1984) *Minds, Brains and Science* (British Broadcasting Corporation, London), p. 102.
Singer, W., Artola, A., Engel, A. K., König, P., and Kreiter, A. K. (1991) Dahlem Conference, Berlin.
Squires, E. J. (1988) *Found. Phys. Lett.* **1**, 13.
Stephan, H., Baron, G., and Frahm, H. D. (1991) *Insectivora* (Springer, Berlin, Heidelberg), pp. 573.
Stapp, H. P. (1993) *Mind, Matter, and Quantum Mechanics* (Springer, Berlin, Heidelberg).
Szentágothai, J. (1978) *Proc. Roy. Soc. London B* **201**, pp. 219–248.
Thorpe, W. H. (1978) *Purpose in a World of Chance. A Biologist's View* (Oxford University Press, Oxford), p. 124.
Ulinski, P. S. (1990) in *Cerebral Cortex*, Vol. 8A, edited by E. G. Jones and A. Peters (Plenum, New York), pp. 139–215.
Ulinski, P. S., and Margoliash, D. (1990) in *Cerebral Cortex*, Vol. 8A, edited by E. G. Jones and A. Peters (Plenum, New York), pp. 217–265.
Walmsley, B., Edwards, F. R., and Tracey, D. J. (1987) *J. Neurosci.* **7**, 1037–1046.

8 The Evolution of Complexity of the Brain with the Emergence of Consciousness

8.1 Introduction

Chapters 6 and 7 describe a considerable advance in understanding the microsites of the neocortex with the manner in which consciousness is experienced. Chapter 8 represents a review of these discoveries before they are incorporated into the final hypothesis of the mind–brain interaction in Chapters 9 and 10. Chapter 8 gives an enlarged and secure basis for these final developments. There is a detailed account of many aspects of exocytosis, which is the functional microunit of the neocortex. It was of particular importance to have an account of the quantal probability of exocytosis.

In the final synthesis it is necessary to have a further account of the experimental studies of mental actions on the brain using non-invasive techniques in conscious human subjects. Finally in this chapter there is further discussion of the postulated units of experience, the psychons, and of their unitary relationship to the neural units, the dendrons.

The key concept of this chapter is that in the evolution of the mammalian brain there had to come to exist in the neocortex levels of complexity in its ultramicrostructure that we may literally call transcendent, because (Eccles 1992 and Chapter 7) they opened the brain to the world of conscious feeling. Before that the living world was *mindless*, as we would now recognize for bacteria, plants, and lower animals. We may ask, how low? The usual answer would be that all mammals, dogs, cats, monkeys, horses, and rats experience feelings and pain and possibly also birds, but not invertebrates and lower vertebrates such as fish and even amphibia and reptiles that have instinctual and learned responses. However, much more experimental testing may be possible in the light of concepts of how animals could use their consciousness as I will mention at the end of the chapter.

We can assume that the mammalian neocortex evolved for the purpose of integrating the greatly increased complexity of the sensory inputs: visual, auditory, tactile, olfactory, gustatory, and proprioceptive, so as to give effective behaviour. We can now try to understand how the functional structure of the mammalian brain could have properties mediating consciousness of another world from that of matter–energy: the world of feelings, thoughts, memories, intentions, emotions.

We have to concentrate on the neocortex because all other parts of the brain such as the striatum, the diencephalon, the cerebellum, and the pons exist in lower vertebrates, reptiles, amphibia, and fish that do not exhibit evidence of conscious feelings (Thorpe 1974, 1978).

8.2 The Mammalian Neocortex

In this enquiry we come to the mammalian neocortex, which is qualitatively similar to ours, though usually much smaller. It has the same neuronal structure as illustrated in Figures 6.3a, b and 8.1a, which show the most numerous and important neurons, the pyramidal cells (B, C, D, E in Figures 6.3a, b and 8.1a); three in Figure 6.3b and five in Figure 8.1a are drawn with many in shadowy outline. The cortex has six layers (Figure 8.1a) and all true pyramidal cells are in laminae V, III, II, each with an apical dendrite and many side branches projecting towards the surface to end as a terminal tuft. A pyramidal cell has a nerve fibre or axon for transmitting information. It projects downwards from the cell, as shown in Figures 8.1a and 8.2a, and leaves the cerebral cortex. It ends eventually in many branches either elsewhere in the cortex or far more distant sites in the brain.

It is necessary to make some further statements with illustrations on the mammalian neocortex before we come to its transcendent properties. It is composed of an immense number, even *thousands of millions*, of individual *nerve cells* that are illustrated in Figures 6.3, 8.1a, and 8.2a, and each is the recipient of information from other nerve cells by means of the fine axonal branches that terminate as *synaptic knobs* or *boutons* (Figures 8.1b and 8.2a). There are thousands of excitatory spine synapses on each pyramidal cell, as are partly illustrated by the spines on the pyramidal cell apical dendrites in Figures 6.3b and 8.1a, b and by the clustered boutons on the apical dendrite and its branches in Figure 8.2a. Even that drawing is inadequate in showing the thousands of boutons on the synaptic spines of the apical dendrites of each pyramidal cell and of course also on other cell dendrites and the cell body.

The bouton (pre) of one synapse is shown in Figure 8.1b making an excitatory synapse across a synaptic cleft (d) on a dendritic spine and containing numerous synaptic vesicles (sv) filled with 5 000 to 10 000 molecules of the specific transmitter substance. Some vesicles are clustered along the synaptic surface of the bouton and they are the principal actors in my story.

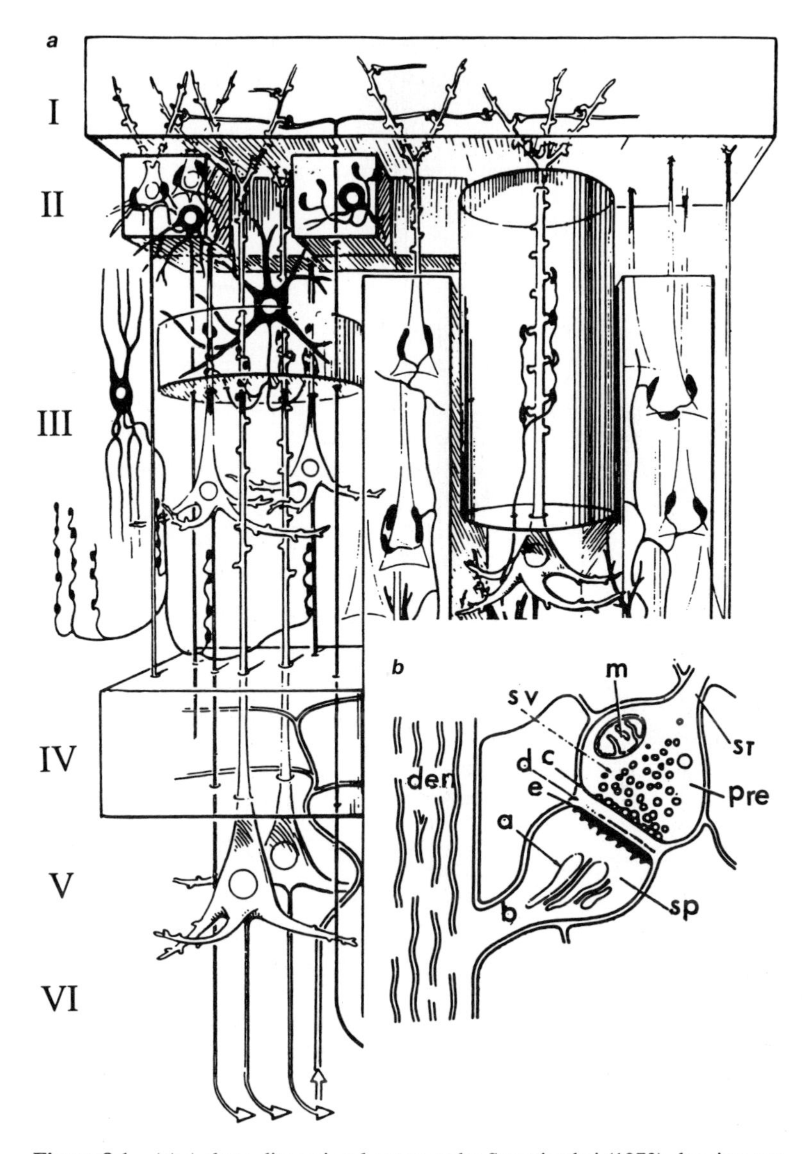

Figure 8.1 (**a**) A three-dimensional construct by Szentágothai (1979) showing cortical neurons of various types. There are two pyramidal cells in lamina V and three in lamina III, one being shown in detail in a column to the right, and two in lamina II. (**b**) The detailed structure of a spine synapse on a dendrite (den. t); st = axon terminating in a synaptic bouton or presynaptic terminal (pre); sv = synaptic vesicles; c = presynaptic vesicular grid; d = synaptic cleft; e = postsynaptic membrane; a = spine apparatus; b = spine stalk; m = mitochondrion (Gray 1982).

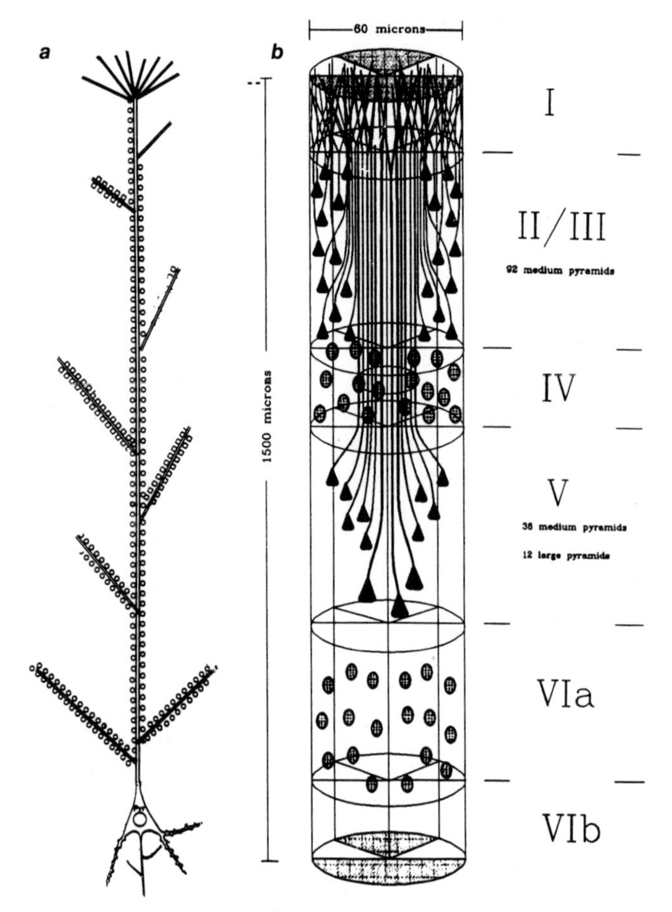

Figure 8.2 (**a**) A drawing of a lamina V pyramidal cell with its apical dendrite showing the side branches and the terminal tuft all studded with spine synapses (not all shown). The soma with its basal dendrites has an axon with axon collateral before leaving the cortex. (**b**) A drawing of the six laminae of the cerebral cortex with the apical dendrites of pyramidal cells of laminae II, III, and V, showing the manner in which they bunch in ascending to lamina I, where they end in tufts. The small pyramids of laminae IV and VI do not participate in this apical bunching (Peters, personal communication, 1991).

Nerve impulses are brief signals of about one millisecond depolarization passing along nerve fibres to finish in the terminal bouton as in Figure 8.1b and as greatly enlarged in Figures 8.3 and 8.4. Transmission across a synapse occurs when an impulse invading the bouton causes a synaptic

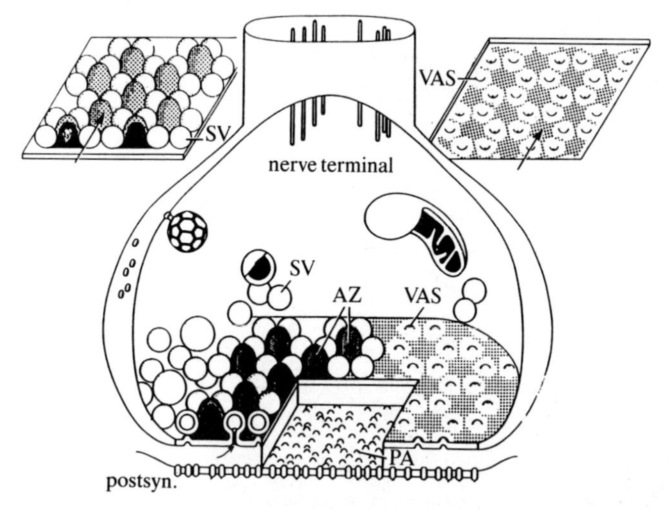

Figure 8.3 The schema of a nerve terminal (bouton) of a mammalian central synapse showing the active zone, the presynaptic vesicular grid with a geometrical design of dense projections (AZ) in triangular array and of synaptic vesicles (SV) in hexagonal array. One vesicle is shown in exocytosis, indicated by an *arrow* in the synaptic cleft. *Below* is the postsynaptic membrane with particle assemblages PA below the cut-out. The presynaptic vesicular grid is stripped off to the right to display the hexagonal arrangement of the presynaptic attachments (VAS). The *inset to the left* shows the presynaptic vesicular grid and the *inset to the right* shows the vesicle attachment sites (VAS). Modified from Akert et al. (1975).

vesicle to discharge its contents of transmitter substance into the synaptic cleft, as indicated by the curved arrow in Figure 8.3, and so to act on the specific receptor sites across the synapse (e in Figure 8.1b and postsyn. in Figure 8.3).

For excitatory synaptic transmission in the cerebral cortex we are specially concerned with glutamate as the transmitter with its action in briefly opening ionic channels that momentarily decrease the electric potential across the postsynaptic membrane, so causing a milli-excitatory postsynaptic potential (EPSP) of the dendrite. By electrotonic transmission along the dendrite there is summation of the milli-EPSPs generated by each bouton activated at about the same time (Figure 4.1). When this occurs for a multitude of boutons (see Figure 8.2a) the summed milli-EPSPs could result in a membrane depolarization of 10–20 mV, which could be enough to generate an impulse in the pyramidal cell that would travel down its axon, shown in Figures 6.3 and 8.1a, eventually to the many synapses of the cerebral cortex on dendrites of neurons or to other regions of the brain.

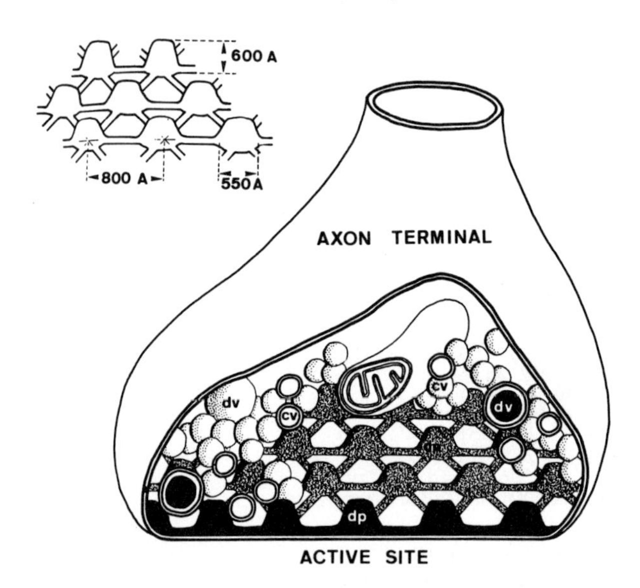

Figure 8.4 An axon terminal, or bouton, showing dense projections (dp) projecting from the active site with cross linkages forming the presynaptic vesicular grid (PVG), which is drawn in the *inset* with dimensions (Pfenninger et al. 1969)

This is the conventional macrooperation of a pyramidal cell of the neocortex, and it can be satisfactorily described by classical physics and neuroscience, even in the most complex design of network theory and neuronal group selection (Szentágothai 1978; Mountcastle 1978; Edelman 1989; Changeux 1985).

It may seem that this generally accepted simplified account of the neuronal mechanism of the neocortex already indicates a high level of complexity in its design. However, this account neglects the conscious feelings that may be generated by the brain activity. In order to move into this field it is necessary to consider in detail the manner of operation of synapses on the pyramidal cells, which is a new level of complexity. Furthermore, these complex neural structures have been postulated to have mental properties, (Edelman 1989; Crick and Koch 1990). For example, Changeux (1985) speaks of 'consciousness being born'. However, Stapp (1991, 1992) asserts that the origin of consciousness cannot be explained by classical physics; quantum physics is necessary. Classical physics is dedicated to matter–energy at all levels of complexity but is not concerned with the mental world. By contrast quantum physics is closely related to the mental world.

So our enquiry into the manner of operation of synapses on pyramidal cells moves into a higher level of complexity in the quest for mind.

8.3 Organization of the Neocortex

There is agreement by Peters and Kara (1987) of Boston and Fleischhauer and Schmolke of Bonn (1984) that the apical dendrites of the pyramidal cells in laminae V, III, and II (Figure 8.1a) bundle together as they ascend to lamina I (Figures 6.8, 6.10, and 8.2b). So they are neural receptor units of the cerebral cortex composed of about 100 apical dendrites plus their branches (Figure 8.2b) that are together called a *dendron*. The enormous synaptic input into the 70–100 apical dendrites bundled into a dendron (Figure 8.2b) can be calculated to be much more than 100 000 synapses, if there are on the average about 2 000 on each apical dendrite (Figure 8.2a).

8.4 The Ultrastructure of Synapses

In Figure 8.1b there is a general drawing of a cortical synapse, but it is now necessary to describe the ultrastructure as revealed to Akert and associates using the techniques of freeze fracture, electron microscopy, and selective staining. It is here that we enter into a higher level of complexity. Figure 8.3 is a key diagram showing as a central feature a nerve terminal, or bouton, confronting the synaptic cleft (d in Figure 8.1b). The inner surface of a bouton can be recognized (c in Figure 8.1b) as an assemblage of synaptic vesicles but is now shown to be a beautiful structure with dense protein projections (DP) in triangular array (AZ in Figure 8.3), forming the presynaptic vesicular grid (PVG) with SVs fitting snugly in hexagonal array (Figure 8.3) and the left inset therein. Figure 8.4 is an interpretative drawing of the dense projections emphasizing the supportive triangularly arranged protein filaments, which are also seen in its inset, with measurements in Ångstroms (1 nm = 10 Å). The spherical SVs, about 40 nm, with their content of transmitter molecules can be seen in the idealized drawings of the PVG (Figure 8.3 and the left inset therein) with the triangularly arranged DPs, AZ, and the hexagonal array of the SVs. The SVs are so intimately related to the presynaptic membrane that it dimples outward to meet them (Figure 8.3, left inset), and when the SVs are stripped off, these dimples reveal the hexagonal pattern of the presynaptic attachment sites (VAS in Figure 8.3 and the right inset therein). The usual number is 40–60 SVs in the single PVG of a bouton (Figures 8.3 and 8.4).

The exquisite design of the PVG can be recognized as having an evolutionary origin for chemically transmitting synapses. In a more primitive form it can be seen in synapses of the mollusc *Aplysia* (Kandel et al. 1987) and the fish Mauthner cell (Korn and Faber 1987). Its essential rationale would appear to be conservation of transmitter molecules during intense synaptic usage.

With chemical synaptic transmission there was not only the problem of manufacturing the transmitter and transporting it to the synaptic site of action, where it was packaged in vesicles, but also the necessity for conservation. As stated above, there are no more than 40–60 vesicles assembled in the PVG ready for liberation in exocytosis. Yet the demand may be caused by presynaptic impulses invading the bouton at about 40 per second. So the necessity for conservation is evident.

8.5 Exocytosis

A nerve impulse propagating into a bouton causes a large influx of Ca^{2+} ions (Figure 8.5b). The input of four Ca^{2+} ions activates a synaptic vesicle via calmodulin and may cause it momentarily to open a channel (Figure 8.5b) through the contacting presynaptic membrane, as indicated by the curved arrow in Figure 8.3, so that its total transmitter content is liberated into the synaptic cleft in a process called *exocytosis*.

At most a nerve impulse evokes a single exocytosis from a PVG (Figure 8.3). This limitation must involve organized complexity of the paracrystalline PVG. It has not yet been explained how exocytosis can be controlled when the nerve impulse causes an influx of Ca^{2+} ions into a bouton that is thousands of times in excess of the four required for the calmodulin that generates one exocytosis (McGeer et al. 1987, p. 96).

Exocytosis is the basic unitary activity of the cerebral cortex. Each all-or-nothing exocytosis of synaptic transmitter results in a brief excitatory postsynaptic depolarization, the EPSP. As already described, summation by electrotonic transmission (Figure 4.1) of many hundreds of these milli-EPSPs is required for an EPSP large enough (10–20 mV) to generate the discharge of an impulse by a pyramidal cell. This impulse will travel along its axon (Figures 6.3, 7.1d, and 7.8) to make effective excitation at its many synapses.

Exocytosis has been intensively studied in the mammalian central nervous system, where it is now possible to move to a new level of complexity by utilizing a single excitatory impulse to generate EPSPs in single neurons that were being studied by intracellular recording. Immense difficulties are

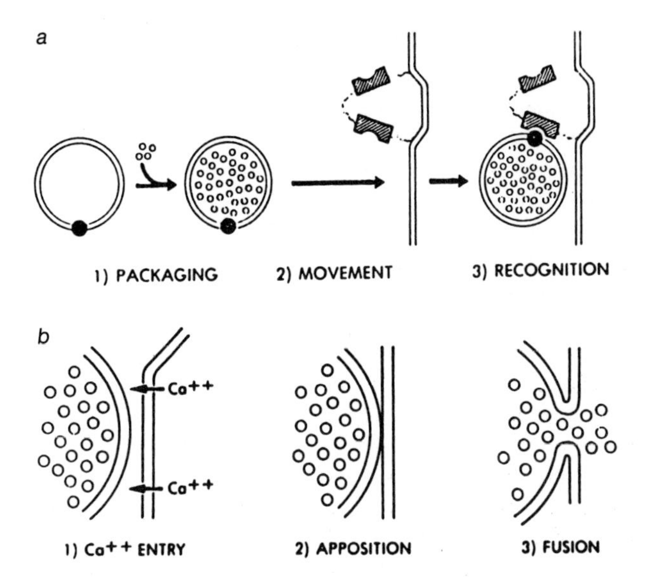

Figure 8.5 Stages of synaptic vesicle development, movement, and exocytosis. (**a**) The three steps involved in filling a vesicle with transmitter and bringing it to attachment to a presynaptic dense projection of triangular shape. (**b**) Stages of exocytosis with release of transmitter into the synaptic cleft, depicting the essential role of Ca^{2+} input (Kelly et al. 1979).

presented by the background noise that was even as large as the signals being studied (Figure 8.6). Fortunately the signal can be repeated many thousands of times for effective averaging above the background noise, and special statistical procedures of deconvolution analysis have been devised to extract the probabilities for exocytoses (Redman 1990).

The initial studies were on the spinal cord, the monosynaptic action on motoneurons by single impulses in the large Ia afferent fibres from muscle (Figure 4.1a)(Jack et al. 1981). Recently it was found (Walmsley et al. 1987) that the signal-to-noise ratio was much better for the neurons projecting up the dorsospinocerebellar tract (DSCT) to the cerebellum, and many quantal responses generated by exocytosis on DSCT neurons were studied. The quantal EPSPs had a mean probability of 0.76.

8.6 The Probability of Exocytosis

Figure 8.6 shows diagrammatically the experimental arrangement for making the most important study of the probability of exocytosis in neurons of the hippocampus, which is a special type of cerebral cortex (Sayer et al. 1989). Advantage was taken of a unique neuronal connection (Figure 8.6a–d). The axon of a CA3 neuron gives off a branch, a Schaffer collateral (Sch), that makes synapses on the apical dendrite of another neuron, type CA1. A microelectrode inserted into a CA3 neuron (A) can set up the discharge of an impulse that goes by the Schaffer collateral to the CA1 neuron to end there and generate an EPSP that is recorded intracellularly (Figure 8.6c). This meticulous technique ensures that the CA1 EPSP is generated by an impulse in a single axon, but as shown in Figure 8.6e, f the EPSPs set up by a stimulus at the arrow below are superimposed on noise that is even larger than the EPSP. Nevertheless, superimposition of many thousands of impulses virtually eliminates the random noise so that smooth EPSPs can be recorded (Figure 8.6g, h) and measured to be about 160 μV (Sayer et al. 1990). Moreover, the statistical technique of deconvolution analysis enables the determination of the probability of release by a nerve impulse of a single synaptic vesicle, an exocytosis. As in simpler situations this probability of release is always less than one, and is in fact very low for the hippocampus with average mean values of 0.27, 0.24, and 0.16 for the three completely reliable experiments on the hippocampus (Redman, personal communication, 1992). So an impulse invading a bouton induces an exocytosis with a probability as low as 1 in 4 to 1 in 6. This is a fundamentally important finding, introducing a new level of complexity.

The control of exocytosis has been investigated by Betz and associates, who have for several years made an intensive study of the proteins of synaptic vesicles in the hope of understanding the quantal release mechanism. The two highly significant proteins are rather similar, synaptophysin and synaptoporin (Marquize-Pouey et al. 1991).

In the dynamic structure of the PVG the dense projections (DP in Figures 8.3 and 8.4) are at least as important as the synaptic vesicles, but there seems to be no study of them complementary to that of synaptic vesicles by Betz and associates. Elucidation of the paracrystalline structure of the PVG requires detailed knowledge of both components. The ultimate goal is to account by quantum physics for the low probability of quantal emissions (exocytoses) in response to nerve impulses invading the bouton.

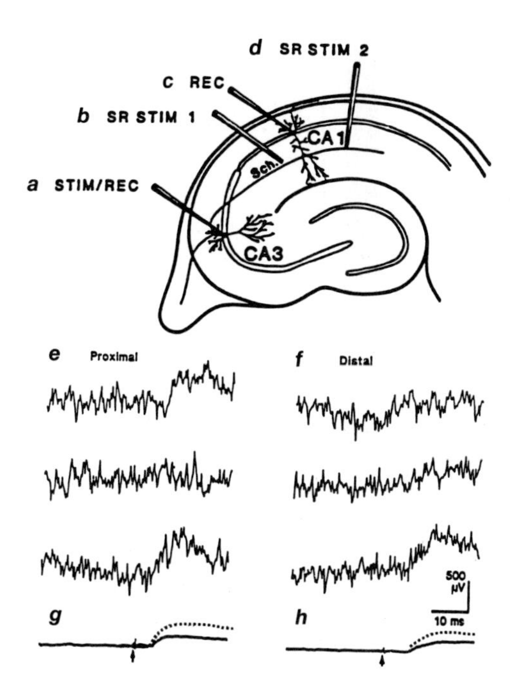

Figure 8.6 (**a–d**) The experimental arrangement for making a probability study of exocytosis of neurons in the *hippocampus*, as described in the text. (**e**) Three records intracellularly recorded as in (c) in response to a stimulus of a single Sch fibre as in (d). (**f**) The average of many thousands of reponses. (**g, h**) Records as in (e, f) but for a Sch fibre distributed more distally to the CA1. In (g) and (h) the *dotted* responses show the increased EPSP produced by an increased exocytosis.

8.7 Psychons

The hypothesis that the dendron is the neural unit of the neocortex leads on to the attempt to discover the complementary mental units which interact with the dendron, for example in intention and attention.

Mental experiences, such as feelings, may not be vague nebulous happenings but instead be microgranular and precisely organized in their immense variety so as to bring about an accurate description of the type of feeling. For example, the experience could be a special sensation from a spot on the right big toe.

The hypothesis has been proposed (Eccles 1990 and Chapter 6) that all mental events and experiences, in fact the whole of the outer and inner sensory experiences, are a composite of elemental or unitary mental experiences

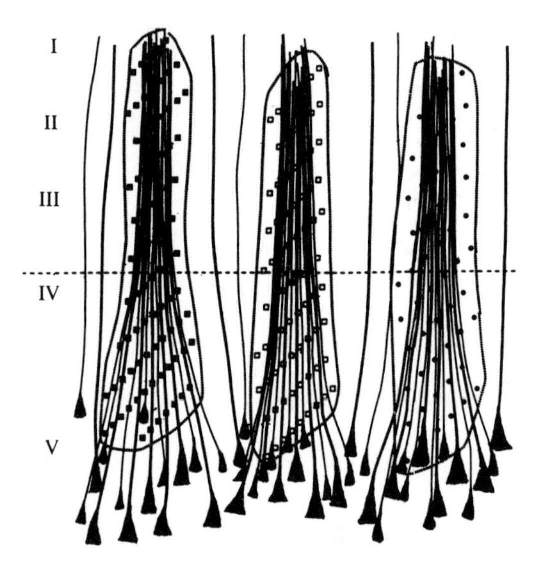

Figure 8.7 A drawing of three dendrons showing the manner in which the apical dendrites of large and medium pyramidal cells bunch together in lamina IV and more superficially, so forming a neural unit. A small proportion of apical dendrites do not join the bunches. The apical dendrites are shown terminating in lamina I. This termination is in tufts that are not shown. The other feature of the diagram is the superposition on each neural unit, or dendron, of a mental unit, or psychon, that has a characteristic marking (*solid squares*, *open squares*, *solid circles*). Each dendron is linked with a psychon giving its own characteristic unitary experience.

at all levels of intensity. Each of these mental units is reciprocally linked in some unitary manner to a dendron, as is illustrated ideally in Figure 8.7 for three dendrons. The three associated mental units are represented as an ensheathing of the three dendrons by designs of solid squares, open squares, and solid circles. Appropriately we name these proposed mental units 'psychons'. According to the unitary hypothesis (Eccles 1990 and Chapter 6) there is a unique linkage of each psychon with its dendron in brain–mind interaction, for example the special feeling in the right big toe.

Psychons are not perceptual paths to experiences. They are the experiences in all their diversity and uniqueness. There could be millions of psychons each linked uniquely to the millions of dendrons. It is hypothesized that it is the very nature of psychons to link together in providing a unified experience.

8.8 Generation of Neural Events by Mental Events

There has been a long history concerning the manner in which voluntary movements can be generated. Some neuroscientists, taking up a materialistic dogma, deny that the non-material mind can effectively influence the brain so as to cause an intended movement. This materialistic dogma neglects the conscious performance which we experience at every moment, even in the linguistic expression of this dogma!

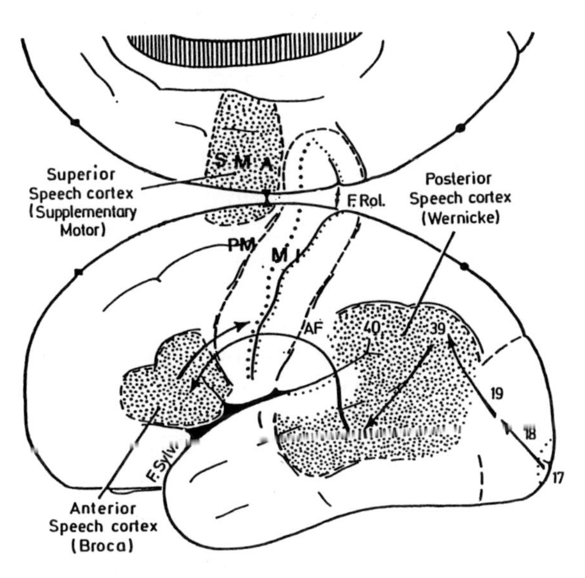

Figure 8.8 The left hemisphere from the lateral side with the frontal lobe to the left. The medial side of the hemisphere is shown as if reflected upwards. F. Rol. is the fissure of Rolando, or the central fissure; F. Sylv. is the fissure of Sylvius. The primary motor cortex, M1, is shown in the precentral cortex just anterior to the central sulcus and extending deeply into it. Anterior to M1 is shown the premotor cortex, PM, with the supplementary motor area, SMA, largely on the medial side of the hemisphere.

Ingvar (1990) introduced the term *pure ideation*, which is defined as cognitive events that are unrelated to any on-going sensory stimulation or motor performances. He stated that

> a study of brain structure activated by pure ideation therefore appears to open up a new approach to understand the human psyche.

Ingvar and his associates at Lund (1965 onwards) introduced the study of the regional cerebral blood flow (rCBF) to display by cerebral ideography the activity of the brain in pure ideation in all the immense variety generated by the psyche. By radio-xenon mapping Roland et al. (1980) demonstrated that in pure motor ideation of complex hand movements there was activation of the supplementary motor area on both sides (SMA, Figures 5.2 and 8.8). By the more accurate technique of PET scanning Raichle and his associates (Figure 10.3) demonstrated a widespread patchy activity of the neocortex during specific *mental operations* in selective attention (Posner et al. 1988).

Thus extensive experimental studies establish that mental intentions (psychons) can effectively activate the cerebral cortex. This increased neural activity can be accounted for if the psychons momentarily caused an increased probability of the exocytoses generated in a bouton by its incoming nerve impulses (Beck and Eccles 1992 and Chapter 9).

The effectiveness of a mental intention causing neural activity has also been well established in the readiness potential (RP) that is recorded by the averaging technique as a slow negative potential over the scalp (Deecke and Lang 1990). It is largest over the supplementary motor area (SMA, Figure 8.8) anterior to the motor cortex (MI in Figure 8.8). By exquisitely designed experiments Libet (1990) has discovered that the readiness potential (RP) begins at least 0.5 s before the subject is conscious (W) of willing the movement, which is at the earliest only 0.2 s before the onset of the movement. So it has been concluded that the brain is active about 0.5 s before the movement is consciously willed (W) by the subject. However, the earlier part of the RP is probably artefactual (Libet 1990, Eccles in General Discussion). So it seems that the *effective willing* (W) of the movement does not occur until 0.2 s before the movement. The mental event of willing (W) can be regarded as preceding the neural events in the brain, particularly in the SMA (Figure 8.8).

The presynaptic vesicular grid provides a unique structure in the attempt to account for the effective action of mental events on the brain by a process that does not infringe the conservation laws of physics (Beck and Eccles 1992 and Chapter 9). A nerve impulse induces an exocytosis in a bouton and an effective milli-EPSP with a mean probability as low as 1 in 5 for the cerebral cortex (Sayer et al. 1989, 1990). This probability requires an explanation in terms of quantum physics. If a mental intention momentarily increased that probability to an average value of 1 in 3, it would have almost doubled the EPSPs for the whole dendron. Thus there would be an *effective mind–brain action* without infringing the conservation laws (Beck and Eccles 1992 and Chapter 9). The very low probabilities of quantal release in the cortex give excellent opportunities for mind–brain interaction in the cerebral cortex. It could be very effective if the psychon's influence were distributed widely to the hundreds of thousands of synapses on a dendron (Figure 8.7). So the low probability of quantal emission in the cerebral cortex was of fundamental significance in the origin of consciousness. The complexity of the ultradesign of cortical synapses presents an ultimate scientific challenge.

As has been described (Eccles 1990, 1992 and Chapters 6 and 7) consciousness gives global experiences from moment to moment of the diverse complexities of cerebral performances, e.g. it would give a mammal global experience of a visual world or a tactual world for guiding its behaviour far beyond what is given by the robotic operations of the visual or tactile cortical areas per se. Thus conscious experiences such as feelings would give evolutionary advantage. This opens up a field of behavioural psychology in which the consciousness of animals could be tested. For example, a reptile such as a tortoise and an insectivore (mammal) such as a hedgehog with its neocortex can be tested to see if the hedgehog displays intelligence when compared to the tortoise, which has no neocortex that could give it consciousness (Eccles 1992 and Chapter 7).

It may seem to be a poor evolutionary design for the fundamental performance of mind–brain interaction because of this dependence on a rather indirect action via quantal probability. It would have been expected that mental intentions would directly excite the neurons of the SMA that are precisely concerned in the intended movement. The indirect action by increase of quantal probability would seem to lack precision and speed, which are of paramount importance in motor control. But in Figure 9.5c movement could occur in 0.2 s. However, once the movement is initiated, it is subject to all the subtle controls of the complex neural machinery of the brain in the conventional neuroscience of motor control.

The hypothesis is that, because the vertebrate brain evolved in the matter–energy world of classical physics, it was mindless, deterministic, and subject to the conservation laws. Then, because of the complex design of the mammalian neocortex with its operation of quantal probability, there came to exist experiences of another world, that of the conscious mind, presumably most primitive and fleeting (Eccles 1992 and Chapter 7). However, with hominid evolution there eventually came higher levels of conscious experiences, and ultimately in *Homo sapiens sapiens*—self-consciousness—which is the unique life-long experience of each human SELF, and which we must regard as a miracle beyond Darwinian evolution (Chapter 10).

8.9 General Summary

Before the evolution of the mammalian brain, it has been proposed in a recent publication (Eccles 1992 and Chapter 7), the animal world was literally mindless, without feelings. In the evolution of even the most primitive mammals, the basal insectivores, there came to exist a neocortex with a higher level of neural complexity, particularly in its pyramidal cell structure (Schmolke 1993). Their apical dendrites have an enormous synaptic input and they form bundles as they ascend through the cortical laminae. There are hundreds of thousands of synaptic inputs, through *boutons*, onto a dendritic bundle, which is the reception unit, named the *dendron*, of the cerebral cortex. The axons of the pyramidal cells have a wide distribution in the brain. In this simplified conventional account of the structure of the cerebral cortex the story of the feelings that may be generated by the brain activity is completely missing.

In developing that story it is necessary to move to a higher level of complexity, the ultramicrosite structure and function of the cortical synapse, as discovered particularly by Akert and associates in Zürich (1975). The boutons of chemical transmitting synapses have a presynaptic ultramicrostructure of a paracrystalline arrangement of dense projections and synaptic vesicles, a *presynaptic vesicular grid.* Its manner of operation in controlling chemical transmission opens up an important field of neural complexity that is still at its inception. The key activity of a synapse concerns a synaptic vesicle that liberates into the synaptic cleft its content of transmitter substance, an exocytosis. There are about 50 synaptic vesicles in the presynaptic vesicular grid. A nerve impulse invading a bouton causes an input of thousands of Ca^{2+} ions, four being necessary to trigger an exocytosis. The fundamental discovery is that at all types of chemical synapses an impulse invading a single presynaptic vesicular grid causes at the most a single exocytosis. There is conservation of the synaptic transmitter by an as yet unknown process of higher complexity.

The conservation challenge becomes intense when it is recognized that synapses of the cerebral cortex have an exceptionally effective conservation with a probability of exocytosis as low as 1 in 5 to 1 in 4 in response to an impulse invading a hippocampal bouton (Figure 8.6b).

Because of the conservation laws of physics, it has been generally believed that non-material mental events can have no effective action on neuronal events in the brain. On the contrary, it has been proposed that all mental experiences have a unitary composition, the units being unique for each type of experience and called *psychons*. It has been further hypothesized that each psychon is linked in a unitary manner to a specific dendron, which is the basis of mind–brain interaction.

Quantum physics gives a new understanding of the mode of operation of the presynaptic vesicular grid and of the probability of exocytosis. Changes in this probability are brought about without an energy input, so the mind could achieve effective action on the brain merely by increasing the probability of exocytosis, for example from 1 in 5 to 1 in 3 (Figure 9.5). That would give a large neuronal response when the mind through its psychons causes this increment in the hundreds of thousands of presynaptic vesicular grids on specific dendrons. A higher level of neural complexity is thus envisaged in order to lead to an understanding of how the mind can effectively influence the brain in conscious volition without infringing the conservation laws.

References to Chapter 8

Akert, K., Peper, K., and Sandri, C. (1975) Structural organization of motor end plate and central synapses, in *Cholinergic Mechanisms*, edited by P. G. Waser (Raven, New York), pp. 43–57.

Beck, F., and Eccles, J. C. (1992) Quantum aspects of brain activity and the role of consciousness, *Proc. Nat. Acad. Sci.* **89**, 11357.

Changeux, J. P. (1985) *Neuronal Man* (Fayard, Paris).

Crick, F., and Koch, C. (1990) Towards a neurobiological theory of consciousness, *Seminars in the Neurosciences* **2**, 263–275.

Deecke, L., and Lang, V. (1990) Movement-related potentials and complex actions: coordinating role of the supplementary motor area, in *The Principles of Design and Operation of the Brain*, edited by J. C Eccles and O. D. Creutzfeldt (Experimental Brain Research, Series 21) (Springer, Berlin, Heidelberg), pp. 303–341.

Eccles, J. C. (1989) *Evolution of the Brain: Creation of the Self* (Routledge, London).

Eccles, J. C. (1990) A unitary hypothesis of mind–brain interaction in the cerebral cortex, *Proc. Roy. Soc. London B* **240**, 433–451.

Eccles, J. C. (1992) The evolution of consciousness, *Proc. Nat. Acad. Sci.* **89**.

Edelman, G. M. (1989) *The Remembered Present: A Biological Theory of Consciousness* (Basic Books, New York).

Gray, E. G. (1982) Rehabilitating the dendritic spine, *Trends Neurosci.* **5**, 5–6.

Ingvar, D. H. (1990) On ideation and 'ideography', in *The Principles of Design and Operation of the Brain*, edited by J. C. Eccles and O. D. Creutzfeldt (Experimental Brain Research, Series 21) (Springer, Berlin, Heidelberg), pp. 433–453.

Jack, J. J. B., Redman, S. J., and Wong, K. (1981) The components of synaptic potentials evoked in cat spinal motoneurones by impulses in single group Ia afferents, *J. Physiol.* **321**, 65–96.

Kelly, R. B., Deutsch, J. W., Carlson, S. S., and Wagner, J. A. (1979) Biochemistry of neurotransmitter release, *Ann. Rev. Neurosci.* **2**, 399–446.

Korn, H., and Faber, D. S. (1987) Regulation and significance of probabilistic release mechanisms at central synapses, in *New Insights into Synaptic Function*, edited by G. M. Edelman, W. E. Gall, and W. M. Cowan (Neuroscience Research Foundation/Wiley, New York), pp. 57–108.

Libet, B. (1990) Cerebral processes that distinguish conscious experience from unconscious mental functions, in *The Principles of Design and Operation of the Brain*, edited by J. C Eccles and O. D. Creutzfeldt (Experimental Brain Research, Series 21) (Springer, Berlin, Heidelberg), pp. 185–205, and General Discussion, pp. 207–211.

McGeer, P. L., Eccles, J. C., and McGeer, E. (1987) *The Molecular Neurobiology of the Mammalian Brain*, 2nd edn (Plenum, New York).

Marquize-Pouey, B., Wisden, W., Malosio, M. L., and Betz, H. (1991) Differential expression of synaptophysin and synaptoporin in RNAs in the post-natal rat central nervous system, *J. Neurosci.* **11**, 3388–3397.

Mountcastle, V. B. (1978) An organizing principle for cerebral function: the unit module and the distributed system, in *The Mindful Brain* (MIT Press, Cambridge MA).

Peters, A., and Kara, D. A. (1987) The neuronal composition of area 17 of the rat visual cortex. IV. The organization of pyramidal cells, *J. Comp. Neurol.* **260**, 573–590.

Pfenninger, K., Sandri, C., Akert, K., and Eugster, C. H. (1969) Contribution to the problem of structural organization of the presynaptic area, *Brain Res.* **12**, 10–18.

Posner, M. I., Petersen, S. E., Fox, P. T., and Raichle, M. E. (1988) Localization of cognitive operations, *Science* **240**, 1627–1631.

Ramón y Cajal, S. R. (1911) *Histologie du Systeme Nerveux* (Maloine, Paris).

Redman, S. J. (1990) Quantal analysis of synaptic potentials in neurons of the central nervous system, *Physiol. Rev.* **70**, 165–198.

Roland, P. E., Larsen, B., Lassen, N. A., and Skinhøj, E. (1980) Supplemental motor area and other cortical areas in organization of voluntary movements in man, *J. Neurophysiol.* **43**, 118–136.

Sayer, R. J., Redman, S. J., and Andersen, P. (1989) Amplitude fluctuations in small EPSPs recorded from CA1 pyramidal cells in the guinea pig hippocampal slice, *J. Neurosci.* **9**, 845–850.

Sayer, R. J., Friedlander, M. J., and Redman, S. J. (1990) The time-course and amplitude of EPSPs evoked at synapses between pairs of CA3/CA1 neurons in the hippocampal slice, *J. Neurosci.* **10**, No. 3, 626–636.

Schmolke, C., and Fleischhauer, K. (1984) Morphological characteristics of neocortical laminae when studied in tangential semi-thin sections through the visual cortex in the rabbit, *Anat. Embryol.* **169**, 125–132.

Stapp, H. P. (1991) Brain–mind connection, *Found. Phys.*, **21**, No. 12.

Stapp, H. P. (1992) in *Nature, Cognition and System*, edited by M. Carvallo (Kluwer, Dordrecht).

Szentágothai, J. (1978) The neuron network of the cerebral cortex: a functional interpretation, *Proc. Roy. Soc. London B* **201**, pp. 219–248.

Szentágothai, J. (1979) Local neuron circuits of the neocortex, in *The Neurosciences Fourth Study Program*, edited by F. O. Schmitt and F. G. Worden (MIT Press, Cambridge MA), pp. 399–415.

Thomas, L., Knaus, P, and Betz, H. (1989) Comparison of the presynaptic vesicle component synaptophysis and gap junction proteins: a clue for neurotransmitter release?, in *Molecular Biology of Neuroreceptors and Ion Channels*, edited by A. Maeliche (Springer, Berlin, Heidelberg).

Thorpe, W. H. (1974) *Animal Nature and Human Nature* (Methuen, London).

Thorpe, W. H. (1978) *Purpose in a World of Chance. A Biologist's View* (Oxford University Press, Oxford).

Walmsley, B., Edwards, F. R., and Tracey, D. J. (1987) The probabilistic nature of synaptic transmission at a mammalian excitatory central synapse, *J. Neurosci.* **7**, 1037–1046.

9 Quantum Aspects of Brain Activity and the Role of Consciousness

F. Beck and J. C. Eccles

9.1 Introduction: The Story of the Eventual Solution

From 1984 onwards I realized that the Margenau story was inadequate. What was needed was a rigorous mathematical treatment that would give precision to the action of mental events on neural events, not just the ill-defined probability field. When I was attached to the Neuroscience Research Program at the Rockefeller University for some months in 1986 and 1987, I organized two mini-symposia in the hope that there could be progress beyond the stage where Margenau left it in *The Miracle of Existence*. Henry Margenau came to both these informal gatherings in New York. Unfortunately Margenau and Wigner were not well enough to make the progress that I hoped for. John Archibald Wheeler was unable to attend, but he nominated two physicists. I have all the documents of these two most disappointing projects. It was evident that the physicists did not realize that here was a great chance for quantum physicists. One physicist insisted that it could all be done by thermal physics, not quantum physics!

I still published on the mind–brain problem, for example, Chapter 7, and several other papers not in this collection, and I completed my evolution book *Evolution of the Brain: Creation of the Self.* I lectured on my theory as presented in Chapters 4, 5, and 6. I was most unhappy all this time, since I was confident that a good quantum physicist could develop a quantum physical solution of the mind–brain problem.

In September 1991 Christoph von Campenhausen and I were together for the fifth time to participate in a two-week summer university course of the *Studienstiftung des deutschen Volkes*, an establishment to promote gifted students of German universities. Our special course was on my evolution book. We were in the Dolomites Conference Centre at Völs. Christoph told me that the most distinguished scientists at the conference were from Darmstadt with Friedrich Beck as the leader.

When I gave the evening lecture in English on my book, it was well understood by the German students, and afterwards I was given a wonderful surprise by Friedrich Beck, whom I had not previously known. He is a quantum physicist and head of the Department of Theoretical Nuclear Physics at the Technical University of Darmstadt. He completely understood the difficulties preventing further progress in the quantum physics of the mind–brain problem. It was at the stage represented by Chapter 6. So we spent a magical hour in the coffee room of the hotel. The next evening we also had an even longer discussion. I had meanwhile given Friedrich what literature I had, and we agreed to cooperate by correspondence, with me providing him with the literature that he would eagerly study.

It was the romantic triumph after my long series of rejections. I had thought that the solution of the mind–brain problem could be accomplished by the Heisenberg uncertainty principle for the minute vesicular aperture allowing exocytosis (Figure 6.13b), which would require a mass of only about 10^{-18} g. Friedrich's calculations showed that the particle involved in the exocytosis had to be orders of magnitude smaller than I had calculated from Margenau's equation. So perforce we turned to the sole remaining possibility, namely that the presynaptic vesicular grid had the exquisite paracrystalline design especially illustrated by Konrad Akert of Zürich (Figure 4.5) with the amazingly controlled exocytosis of never more than one. A nerve impulse causes a large influx of Ca^{2+} ions into a bouton, thousands of times more than the four Ca^{2+} ions required for exocytosis (Figure 4.6). So deficiency of Ca^{2+} input is not the cause of the invariably observed conservation reaction with an exocytosis of never more than one for an impulse into a bouton. Usually it was one in three or one in four. Beck immediately realized that a key problem was to account for this low probability of exocytosis by using quantum physics. He realized that quantum physics could account for this and that, if the mental influence could momentarily increase probability, the mind could effectively act on the brain without infringing the conservation laws.

Meanwhile, he was studying the proposed solutions of the mind–brain problem by quantum physicists and by neuroscientists, as described in Chapter 3, and as stated in our paper (Beck and Eccles 1992). All previous attempts had been treating the cerebral cortex in very general terms. There was no reference to the presynaptic vesicular grid or to the bundling of the apical dendrites of the pyramidal cells to give the intense receptor property for both synaptic and psychon inputs.

The rest of this chapter, co-written with Beck (Beck and Eccles 1992), appeared in the *Proceedings of the National Academy of Sciences* at the end of 1992. We put forward the key hypothesis that mental intention of the self becomes neurally effective by *momentarily increasing the probabilities for exocytoses* in a whole dendron and in this way couples the large number of probability amplitudes to produce coherent action (Figures 6.10 and 9.2).

Our hypothesis offers a natural explanation for voluntary movements caused by mental intentions without violating physical conservation laws. It has been shown experimentally that intentions and attentions activate the cerebral cortex in certain well-defined regions prior to the movement (Figures 5.2 and 5.4).

The unitary hypothesis transforms the manner of operation of an intention. One has to recognize that, in a lifetime of learning, the intention to carry out a particular movement would be directed by the self to those particular dendrons of the neocortex that are appropriate for bringing about the required actions. We believe that the proposed hypothesis accounts for action across the mind–brain interface.

This chapter represents the culmination of my life-long quest to find the scientific explanation for dualism. When examined critically it will be recognized that experiments have not been done to demonstrate that mental events can increase the probability of exocytosis and so effectively increase the neural activities of specific cortical areas. Such a probability study as illustrated in Figure 8.6 is extremely difficult to carry out on cortical slices of the guinea pig hippocampus, a specialized type of cerebral cortex. It is quite impossible ethically to carry out the experiments of Figure 8.6 on human brains. A more feasible test could be to demonstrate conditioning by various physiological and pharmacological treatments of the Ia synapses on motoneurons whose exocytotic probabilities were being measured (Jack et al. 1981a, 1981b). There was the enhancing effect by a prior stimulus, probably owing to the residual Ca^{2+} in the bouton, and by 4-aminopyridine (Chapter 4.5) and the depressant effect of high-frequency stimulation, halving at about 33 Hz as described in Chapter 4.4. Thus the observed exocytotic probability is labile and so presumably open to modification by the self, as indicated in Figure 5.4.

(J. C. E.)

9.2 Quantal Selection of Bouton Exocytosis

There has been increasing interest in recent years in the relation between quantum mechanics, brain activity, and consciousness (Donald 1990; Margenau 1984; Squires 1988; Stapp 1991). On the part of quantum physics the impetus came from the interpretation of the measurement process, which is still being debated, even after more than sixty years of overwhelmingly successful applications of quantum theory. However, there is the question in neuroscience (Eccles 1990 and Chapter 6) whether quantum action is needed to understand the functioning of the brain in its subtle relations to experience, memory, and consciousness.

It was Wigner (1967), in his stringent analysis of the consequences of measurements in a Stern–Gerlach experiment, who first speculated that the von Neumann collapse of the wave function (von Neumann 1955) actually occurs through an act of consciousness in the human brain and is not describable in terms of ordinary quantum mechanics. It is well known that the 'reduction of the wave function' which occurs in the measurement process does not obey the causality principle, in contrast to the continuous time evolution governed by the Schrödinger equation. This is most drastically demonstrated in the Einstein–Podolsky–Rosen experiment (Einstein, Rosen, and Podolsky 1935). The wave function as a probability amplitude is not a material field. Its only conservation law is the conservation of probability. These facts offer the fascinating possibility of *different final states* as outcomes of *identical dynamical processes* without change in the initial conditions or in the external control parameters, such as the energy input. Quantal selection in this way provides an *interface* between the physiological processes in the brain and the non-deterministic action of the mind.

More recently similar ideas to the ones presented by Wigner were put forward and partly combined with Everett's 'many-worlds' interpretation (Everett 1957) of quantum theory (Squires 1988). Other authors (e.g. Stapp 1989) have related quantum theory to consciousness on the basis of the usual interpretation of the state vector as a superposition of actualities (or 'propensities' in Popper's nomenclature (Popper 1990; Popper and Eccles 1977). All these attempts have in common that they argue in very general terms with respect to an interpretation of quantum mechanics and its probability concepts. Not much connection has been made, however, to the empirically established facts of brain physiology, nor have the authors attempted to locate a quantal process in the functional microsites of the neocortex.

In this work we contribute to filling this gap by putting forward a quantum mechanical description of bouton exocytosis. The next section gives an outline of the structure and activity of the neocortex as an introduction to the quantum mechanical model. This is set up in the subsequent section. Finally the hypothesis for a coherent coupling of the individual probability amplitudes is introduced, allowing for the influence of consciousness and leading to an amplification of the summed postsynaptic potentials. In this model, action becomes possible by influence of a conscious will.

9.3 Neocortical Activity: Bouton Exocytosis

Figure 9.1a illustrates the universally accepted six laminae of the neocortex (Szentágothai 1978) with two large pyramidal cells in lamina V, three in lamina III, and two in lamina II. The pyramidal apical dendrites finish in a tuft-like branching in lamina I (Figure 9.2a). There is agreement by Peters and Fleischhauer and their associates (Peters and Kara 1987; Schmolke and Fleishhauer 1984) that the apical bundles or clusters diagrammatically shown in Figure 9.2b are the basic anatomical units of the neocortex. They are observed in all areas of the cortex that have been investigated and in all mammals including humans. Approximate values can be given for the synaptic connectivity of an apical bundle. The input would be largely by the spine synapses (Figures 9.1 and 9.2a), which would number over 5 000 on a lamina V apical dendrite with its lateral branches and terminal tuft (Figure 9.2a), but more usually there would be about 2 000. If there are 70–100 apical dendrites in a bundle, the total spine synapses would number well over 100 000. It has been proposed that these bundles are the cortical units for reception (Eccles 1990 and Chapter 6), which would give them a preeminent role. Since they are composed essentially of dendrites, the name *dendron* was adopted.

Figure 9.1b illustrates a typical spine synapse that makes an intimate contact with an apical dendrite of a pyramidal cell (Figures 9.1 and 9.3a, d). The ultrastructure of such a synapse has been extensively studied by Akert and his associates (Akert, Peper, and Sandri 1975; Pfenninger, Sandri, Akert, and Eugster 1969). From the inner surface of a bouton confronting the synaptic cleft (d in Figure 9.1b, the active site in Figure 9.3a), dense projections in triangular array form the presynaptic vesicular grid (PVG) (Figure 9.3a–e). Figure 9.3b is a photomicrograph of a tangential section of a PVG showing the dense projections in triangular array with the faint synaptic vesicles fitting snugly in hexagonal array. The spherical synaptic vesicles, about 40 nm in diameter, with their content of transmitter molecules, can be seen in the idealized drawings of the PVG (Figures 9.3c, d) with the triangularly arranged dense projections, the active zone, and the hexagonal array of synaptic vesicles (Akert, Peper, and Sandri 1975; Pfenninger, Sandri, Akert, and Eugster 1969).

The exquisite design of the PVG can be recognized as having an evolutionary origin for chemically transmitting synapses. In a more primitive form it can be seen in synapses of the fish Mauthner cell (Korn and Faber 1987). Its essential rationale can be recognized as a conservation of transmitter molecules during intense synaptic usage.

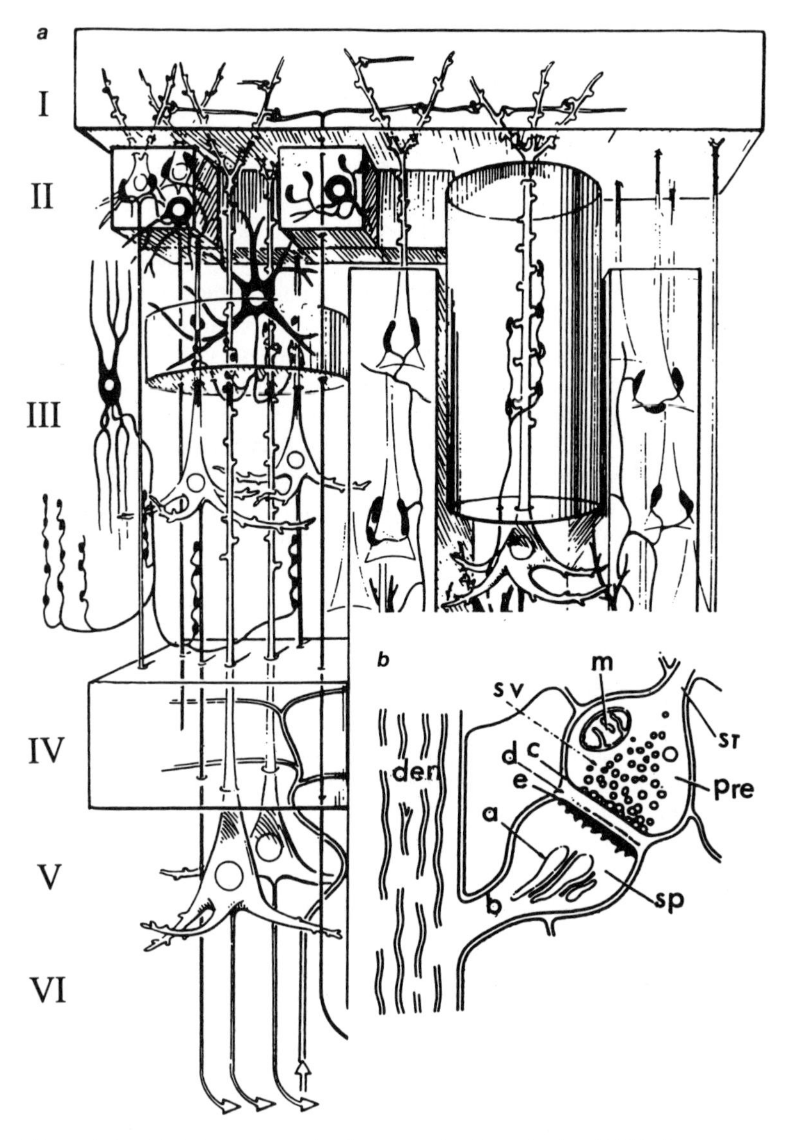

Figure 9.1 (**a**) A three-dimensional construct by Szentágothai (1978) showing cortical neurons of various types. There are two pyramidal cells in lamina V and three in lamina III, one being shown in detail in a *column to the right*, and two in lamina II. (**b**) The detailed structure of a spine (sp) synapse on a dendrite (den); st = axon terminating in synaptic bouton or presynaptic terminal (pre); sv = synaptic vesicles; c = presynaptic vesicular grid (PVG); d = synaptic cleft; e = postsynaptic membrane; a = spine apparatus; b = spine stalk; m = mitochondrion (Gray 1982).

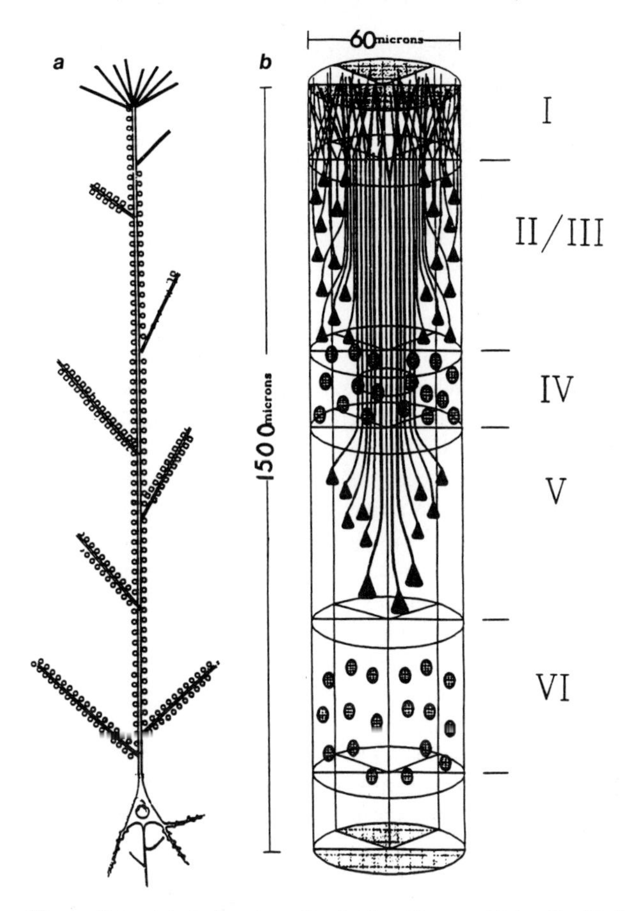

Figure 9.2 (**a**) A drawing of a lamina V pyramidal cell with its apical dendrite showing the side branches and the terminal tuft, all studded with spine synapses (not all shown). The soma with its basal dendrites has an axon with axon collateral before leaving the cortex. (**b**) The six laminae of the cerebral cortex with the apical dendrites of pyramidal cells of laminae II, III, and V, showing the manner in which they bunch in ascending to lamina I, where they end in tufts. The small pyramids of laminae IV and VI do not participate in this apical bunching (A. Peters, personal communication, 1992).

A nerve impulse propagating into a bouton causes a large influx of Ca^{2+} ions (Figure 8.5). The input of four Ca^{2+} ions to a synaptic vesicle may cause it momentarily to open a channel through the contacting presynaptic membrane so that its total transmitter content is liberated into the synaptic cleft (d in Figure 9.1b and Figure 9.3f, g) in a process called *exocytosis*.

At most a nerve impulse evokes a single exocytosis from a PVG (Figure 9.3f, g). This limitation is probably due to the vesicles' being embedded in the paracrystalline PVG (Figure 9.3b–d).

Exocytosis is the basic unitary activity of the cerebral cortex. Each all-or-nothing exocytosis of synaptic transmitter results in a brief excitatory postsynaptic depolarization (EPSP). Summation by electrotonic transmission of many hundreds of these milli-EPSPs is required for an EPSP large enough (10–20 mV) to generate the discharge of an impulse by a pyramidal cell. This impulse will travel along its axon (Figure 9.2a) to make effective excitation at its many synapses. This is the conventional macrooperation of a pyramidal cell of the neocortex (Figure 9.1) and it can be satisfactorily described by classical physics and neuroscience, even in the most complex design of network theory and neuronal group selection (Edelman 1989; Mountcastle 1978; Szentágothai 1978).

Exocytosis has been extensively studied in the mammalian central nervous system where it is possible to refine the study by utilizing a single excitatory impulse to generate EPSPs in single neurons that are being studied by intracellular recording. The initial studies were on the monosynaptic action on motoneurons by single impulses in the large Ia afferent fibres from muscle (Jack, Redman, and Wong 1981). More recently (Walmsley, Edwards, and Tracey 1987) it was found that the signal-to-noise ratio was much better for the neurons projecting up the dorso-spino-cerebellar tract (DSCT) to the cerebellum.

This successful quantal resolution for DSCT neurons and motoneurons gives confidence in the much more difficult analysis for neurons of the cerebral cortex, which provide the key structures for neural events that could be influenced by mental events. The signal-to-noise ratio was so low in the studies of CA1 neurons of the hippocampus that so far only three quantal analyses have been reliable in the complex deconvolution procedure (Figure 8.6).

In the most reliable of these, a single axon of a CA3 hippocampal pyramidal cell set up an EPSP of quantal size 278 μV (mean value) in a single CA1 hippocampal pyramidal cell with approximately equal probabilities of release at each active site ($n = 5$) of 0.27 (Sayer, Friedlander, and Redman 1990). In the alternative procedure the single CA3 impulse projecting to a CA1 pyramidal cell was directly stimulated in the stratum radiatum. The EPSPs delivered by the deconvolution analysis of two CA1 pyramidal cells were of quantal sizes 224 μV and 193 μV with probabilities of 0.24 ($n = 3$) and 0.16 ($n = 6$), respectively (Sayer, Redman, and Andersen 1989). For a systematic review, see Redman (1990).

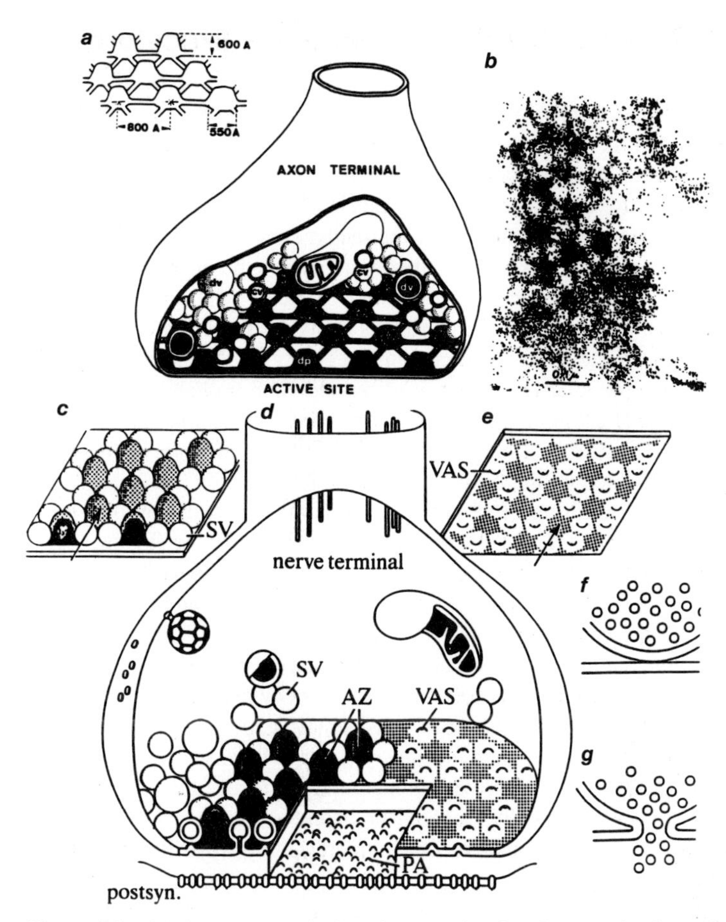

Figure 9.3 (**a**) An axon terminal, or bouton, showing dense projections (dp) projecting from the active site with cross linkages forming the PVG, which is drawn in the *inset* with dimensions (Pfenninger, Sandri, Akert, and Eugster 1969). (**b**) A tangential section through the presynaptic area. A section of the patterns of dense projections and synaptic vesicles, triangular and hexagonal, of the PVG is clearly represented. (*Bar* = 0.1 μm.) (Reprinted from Akert, Peper, and Sandri 1975 with permission.) (**c–e**) The active zone (AZ) of the mammalian central synapse showing the geometrical design (Gray 1982). SV = synaptic vesicle; VAS = vesicle attachment site; PA = postsynaptic area. (**f**) Synaptic vesicle in apposition. (**g**) Exocytosis (Kelly, Deutsch, Carlson, and Wagner 1979).

9.4 A Quantum Mechanical Model of Exocytosis

In the whole electrophysiological process of building up the summed EPSP experienced in the soma, there is only one element where quantal processes can play a role. When a bouton is activated by a nerve impulse, exocytosis occurs only with a certain probability which is much less than one. This calls for, in principle, the introduction of thermodynamic or quantum statistical concepts. We wish to make it clear that we adhere to the quantal standpoint. As has been pointed out by several authors (Eccles 1990 and Chapter 6; Squires 1988; Stapp 1991), the conscious action of the brain could hardly be understood if the brain functioned in its entirety on the basis of classical physics. Also the relative constancy of the emission probability for single bouton release (Redman 1990; Sayer, Friedlander, and Redman 1990; Sayer, Redman, and Andersen 1989; Walmsley, Edwards, and Tracey 1987) can hardly be explained on the basis of thermal fluctuations.

Because the resulting EPSP is the *independent statistical sum* of several thousands of local EPSPs at spine synapses on each dendrite (Figure 9.2a), we can concentrate on the process of exocytosis at each individual bouton.

Exocytosis is the opening of a channel in the PVG and discharge of the vesicle's transmitter molecules into the synaptic cleft (Figure 9.3f, g). It is, as a whole, certainly a classical membrane-mechanical process. Though the detailed molecular processes that cause exocytosis are not yet fully understood, there has recently been considerable experimental progress in analysing the membrane fusion machinery (for a review, see Jessel, Kandel, Lewin, and Reid 1993). There exists increasing evidence that the complex process of exocytosis and its probabilistic nature are governed by a *trigger mechanism*, which may involve quantum transitions between metastable molecular states.

To investigate further the possible role of quantum mechanics in the probabilistic discharge, one has to set up a model for the trigger mechanism by which Ca^{2+} prepares the vesicles of PVG for exocytosis. For this purpose we adopt the following concept: preparation for exocytosis means bringing the paracrystalline PVG into a metastable state from which exocytosis can occur. The trigger mechanism is then modelled by the motion of a quasi-particle with one degree of freedom along a collective coordinate, and over an activation barrier (Figure 9.4). This motion sets in by a quantum mechanical tunnelling process through the barrier (similar to radioactive decay).

The question whether a microsystem behaves classically or has to be treated by quantum physics is not only a question of its size, but also a

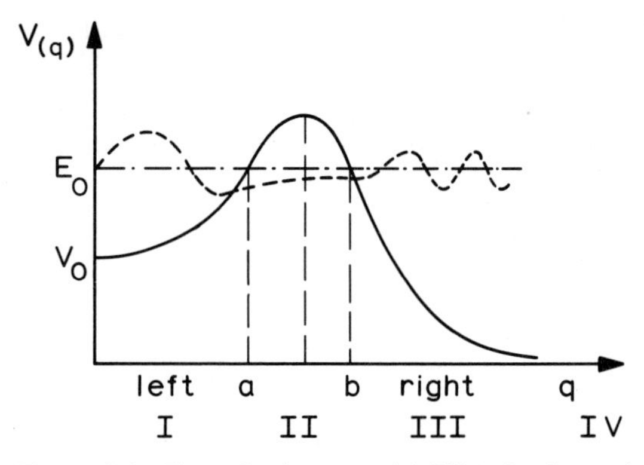

Figure 9.4 The collective potential, $V(q)$, for the motion of the quasi-particle of energy E_0 which triggers exocytosis. The *dashed curve* sketches a tunnelling state through the barrier. At the beginning the wave packet is located in the *left-hand* area. After a time τ the amplitude has a *left-hand* part and a *right-hand* part from which follow the probabilities for the occurrence of exocytosis. a and b = classical turning points; V_0 = potential at the origin of the collective coordinate, which is characterized by the instable situation in the PVG after excitation by a nerve impulse.

question of its embedding in larger surroundings at finite temperature. The situation is determined by two characteristic energies:

(i) The thermal energy, E_{th}, which the quasi-particle adopts in thermal surroundings of temperature T. For our quasi-particle with one degree of freedom this is given by

$$E_{\text{th}} = \tfrac{1}{2} k_{\text{B}} T, \tag{9.1}$$

where k_{B} is the Boltzmann constant.

(ii) The quantum mechanical zero-point energy, E_0, of a particle of mass M localized over a distance of Δq. This follows from Heisenberg's uncertainty relation:

$$\Delta p \Delta q \geq 2\pi\hbar. \tag{9.2}$$

If we adopt the lower limit (this would correspond to the ground state in a confining potential well) we obtain, by identifying the momentum uncertainty, Δp, with its corresponding kinetic energy,

$$E_0 \approx \frac{(\Delta p)^2}{2M} = \left(\frac{2\pi\hbar}{\Delta q}\right)^2 \frac{1}{2M}. \tag{9.3}$$

We can now define a borderline between the quantal and the classical regime by putting

$$E_0 = E_{\text{th}}. \tag{9.4}$$

$E_0 \gg E_{th}$ is then the *quantal* regime and $E_{th} \gg E_0$ the *thermal* regime.

For fixed T and Δq Equation 9.4 determines a critical mass, M_c, for the quasi-particle. Dynamic processes which involve quasi-particle masses much larger than M_c belong to the classical regime, whereas for $M \ll M_c$ we are in the quantal regime. Taking T = 300 K and $\Delta q \approx 1$ Å, we obtain

$$M_c \approx 10^{-23}\,\text{g} \approx 6 M_H, \tag{9.5}$$

where M_H is the mass of the hydrogen atom.

This estimate shows quite clearly that a quantum mechanical trigger for exocytosis must reside in a molecular process, for example, the movement of a hydrogen bridge by electronic rearrangement.

To make the model quantitative we attribute to the triggering process of exocytosis a continuous collective variable, q, for the quasi-particle. The motion is characterized by a potential energy, $V(q)$, which may take on a positive value at stage I (Figure 9.4), according to the metastable situation before exocytosis, then rises towards a maximum at stage II, and finally drops to zero (arbitrary normalization) at stage IV. Figure 9.4 gives a qualitative sketch of this potential barrier.

The time-dependent trigger process of exocytosis is described by the one-dimensional Schrödinger equation for the wave function $\psi(q;t)$,

$$i\hbar \frac{\partial}{\partial t}\psi(q;t) = -\frac{\hbar^2}{2M}\frac{\partial^2}{\partial q^2}\psi(q;t) + V(q)\psi(q;t). \tag{9.6}$$

The initial condition for t = 0 (stage I, beginning of exocytosis) is a wave packet left of the potential barrier (Figure 9.4).

The solution of Equation 9.6 gives the wave function after time t, which consists of part of the wave packet still residing to the left of the barrier while another part has penetrated into the space to the right (see Figure 9.4). The quantities

$$p_1(t) = \int_{\text{left}} \psi^*(q;t)\,\psi(q;t)\,\mathrm{d}q, \tag{9.7a}$$

$$p_2(t) = \int_{\text{right}} \psi^*(q;t)\,\psi(q;t)\,\mathrm{d}q \tag{9.7b}$$

are the time-dependent probabilities that exocytosis will not occur (p_1) or that exocytosis will occur (by action of the trigger) (p_2). Of course, from normalization

$$p_1(t) + p_2(t) = 1, \tag{9.8}$$

and this is the conservation law which the probabilities have to obey.

The two parts of the wave packet at time t, left and right of the barrier (Figure 9.4), are the two possibilities (or states) for which the individual process can make a selection *without any predetermined fixation*. Selecting one state or the other is the acausal process which von Neumann calls 'reduction of the wave function' (Von Neumann 1955). The statistical probabilities of Equation 9.7 determine the frequencies of many repetitions of identical processes. For the individual event they only determine a *tendency* (or 'propensity' (Popper 1990)) for selecting one state or the other.

The relevant time τ has to be defined as the duration of the quasi-stable situation in the paracrystalline PVG, after which the metastable state changes into a stable one where no exocytosis is possible. In our model this can be visualized as a dropping of the potential left of the barrier to a value for which barrier penetration has vanishingly small probability.

An approximate solution of the barrier penetration problem can be obtained from the Wentzel–Kramers–Brillouin (WKB) method (Messiah 1961). According to this method the transmission coefficient T of a particle of mass M and energy E through the barrier is given by

$$T = \exp\left\{-2\int_a^b \frac{\sqrt{2M[V(q)-E]}}{\hbar}\,\mathrm{d}q\right\}, \tag{9.9}$$

where a and b denote the classical turning points for the motion left and right of the barrier (Figure 9.4). The probability per unit time for penetration of the barrier, w, can be obtained by the number of attempts which the particle undertakes to reach the barrier, ω_0, times the barrier transmission coefficient T:

$$w = \frac{\mathrm{d}p}{\mathrm{d}t} = \omega_0 T, \tag{9.10}$$

where $\hbar\omega_0 = E_0$, by appeal to Bohr's correspondence principle. Combining this with the general considerations formulated after Equation 9.7, we obtain

$$p_2(\tau) = \tau w = \tau\omega_0 T. \tag{9.11}$$

At this stage we can estimate some numbers. Let us assume that the energy $E = E_0$ of the initial state (the wave packet localized to the left of the barrier) is the zero-point energy of a wave packet localized over the dimension of the atomic site of the trigger, which we estimate roughly to be $\Delta q \approx 1\,\text{Å}$, and let us take for the mass M of our quasi-particle the mass of one hydrogen atom, $M = 1.7 \times 10^{-24}\,\text{g}$. Then we obtain, using Equation 9.3, $E_0 \approx 8.3 \times 10^{-2}\,\text{eV}$. This leads to a frequency, from $E_0 = \hbar\omega_0$, of $\omega_0 \approx 1.3 \times 10^{14}\,\text{s}^{-1}$. Estimating τ, the time of the metastable instability, to be of the order of electronic transition processes, that is, $\tau \approx 10^{-13}$–$10^{-14}\,\text{s}$, we obtain from Equation 9.11 $p_2(\tau) \approx 10T$ to $100T$. With the observed p_2 of about 0.25 this gives for the barrier penetration factor T the reasonable span of 4×10^{-2} to 4×10^{-3}. From these numbers we conclude that the trigger process for exocytosis belongs to the *femtosecond regime* of membrane quantum chemistry.

Up to now we have based our model considerations on the release of a single vesicle located in the vesicular grid of the bouton. As was pointed out in the previous section there are about 40 vesicles altogether in this paracrystalline structure, but never does more than one vesicle emit transmitter molecules into the synaptic cleft after stimulation by a nerve impulse. This certainly means that the vesicles in the vesicular grid do not act independently, but rather that *immediately* after one vesicle is triggered for releasing its content the interaction between them blocks further exocytosis. The paracrystalline structure of the PVG makes it possible to have long range interactions between the constituents, as is well known from ordered quantum systems. According to our numerical estimates, the relaxation time for the blocking process is of the order of femtoseconds.

With this observation we can discuss the many-body aspect of exocytosis from the vesicular grid. To this end we attribute schematically to each vesicle in the grid two states, ψ_0 and ψ_1, where ψ_0 is the state *before* and ψ_1 the state *after* exocytosis has been triggered. We can safely assume that the different vesicles are so well separated that we can treat them as distinguishable particles. The wave function of N vesicles is then a product of denumerable states:

$$\Psi(1, \ldots, N) = \psi_{i_1}^{(1)} \cdot \psi_{i_2}^{(2)} \cdots \psi_{i_N}^{(N)}; \qquad i_j = \{0, 1\}. \tag{9.12}$$

Before exocytosis the wave function has the form

$$\Psi_0(1, \ldots, N) = \psi_0^{(1)} \cdots \psi_0^{(N)}. \tag{9.13}$$

The observation that in response to a presynaptic impulse only one vesicle can empty its transmitter molecules into the synaptic cleft leads to a properly normalized wave function after the trigger for exocytosis has acted of the form

$$\begin{aligned}\Psi_1(1, \ldots, N) = & \frac{1}{\sqrt{N}} \big(\psi_1^{(1)} \cdot \psi_0^{(2)} \cdots \psi_0^{(N)} + \psi_0^{(1)} \cdot \psi_1^{(2)} \cdot \psi_0^{(3)} \cdots \psi_0^{(N)} \\ & + \cdots + \psi_0^{(1)} \cdots \psi_0^{(N-1)} \cdot \psi_1^{(N)} \big).\end{aligned} \tag{9.14}$$

All other states are pushed up in energy by the long range interaction so far that they cannot be excited by the nerve impulse.

Calculating the probability for exocytosis from the N-body wave functions (Equations 9.13 and 9.14) one obtains the same result as was obtained from the barrier penetration problem of one vesicle, since the trigger can change the state of only one vesicle, but there are N such possiblities. This leads to the observable consequence that the probability for exocytosis of one bouton does not depend on the number of identical vesicles occupying the PVG.

In this section we have set up a quantum mechanical model for exocytosis of boutons after stimulation by a nerve impulse. We linked our description as closely as possible with the known functional structure of dendrites and their synapses. The model can describe with reasonable parameter inputs the quantal bouton release. It introduces into the activity of the neocortex a quantum probabilistic aspect which leads to a selection of choices according to a quantum probability amplitude. This will in turn be used in the next section to postulate coherent couplings of these probability fields in order to produce action influenced by consciousness.

9.5 Generation of Neural Events by Mental Events

Ingvar (1990) introduced the term *pure ideation*, which is defined as cognitive events that are unrelated to any on-going sensory stimulation or motor performances. He stated that

> a study of brain structures activated by pure ideation therefore appears to open up a new approach to understanding the human psyche.

He and associates at the University of Lund introduced the study of the regional cerebral blood flow to display by cerebral ideography the activity of the brain in pure ideation in all the immense variety generated by the psyche. By radio-xenon mapping Roland et al. (1980) (Figure 5.2) demonstrated that in pure motor ideation of complex hand movements there was activation of the supplementary motor area (SMA) on both sides. By the more accurate technique of positron emission tomography (PET scanning) Raichle and his associates (Posner, Petersen, Fox, and Raichle 1988) demonstrated a widespread patchy activity of the neocortex during specific mental operations in selective attention.

In a complementary procedure, averaging techniques have been used to record the electrical fields apparently generated by the brain (centred on the SMA) in the willing of a movement, the so-called readiness potential (Deecke and Lang 1990). In exquisitely designed experiments Libet (1990) has discovered that conscious willing occurs about 200 ms before the movement (see Libet 1990, General Discussion).

So recent experimental studies establish that mental intentions to move can effectively activate the cerebral cortex; hence we must return to the microsite studies of neocortical activity.

Intracellular recording from a cortical neuron, a hippocampal pyramidal cell, discloses a continuous intense activity (Figure 8.6e, g) which can be interpreted as milli-EPSPs generated by the continuous synaptic bombardment with exocytoses by the thousands of boutons on its dendrites (Figure 9.2a). Deconvolution analysis has shown that an impulse invading a bouton (Figure 9.3a, d) evokes an exocytosis and a milli-EPSP with a probability of about 0.2–0.3 (Sayer, Friedlander, and Redman 1990; Sayer, Redman, and Andersen 1989).

Combining these observations with our quantum mechanical analysis of bouton exocytosis, we present now the hypothesis that the mental intention (the volition) becomes neurally effective by *momentarily increasing the probability of exocytosis* in selected cortical areas such as the supplementary motor area neurons (Eccles 1982). In the language of quantum mechanics this means a *selection of events* (the event that the trigger mechanism has functioned, which is already prepared with a certain probability, cf. Figure 9.4 and Equation 9.7). This act of selection is related to Wigner's selection process of the mind on quantal states (Wigner 1967), and its mechanism clearly lies beyond ordinary quantum mechanics. This selection mechanism effectively increases the probability for exocytosis, and in this way generates increased EPSPs *without violation of the conservation laws*. As has been pointed out in Chapter 9.3 and demonstrated in Chapter 9.4, *quantum selection is the only possible way* of producing different final states from identical initial conditions in identical dynamical situations, and thus with the same values of the conserved quantities. Such a situation could not prevail in a purely classical process, where a change in the final state necessarily implies a change either of the initial conditions or of the dynamics. Even in the recently extensively discussed processes governed by classical 'deterministic chaos', the final outcome is predetermined by the initial conditions, though in an extremely sensitive manner. Classical chaotic motion is characterized by extreme instabilities with respect to small changes and cannot therefore account for regular brain processes such as exocytosis. [1]

Furthermore, the interaction of mental events with the quantum probability amplitudes for exocytosis introduces a coherent coupling of a large number of individual amplitudes of the hundreds of thousands of boutons in a dendron. This then leads to an overwhelming variety of actualities, or modes, in the activity of every micro-unit of the neocortex. Physicists

---

[1] Chaotic processes may play a role in pathological situations.

will realize the close analogy to laser action, or, more generally, to the phenomenon of self-organization.

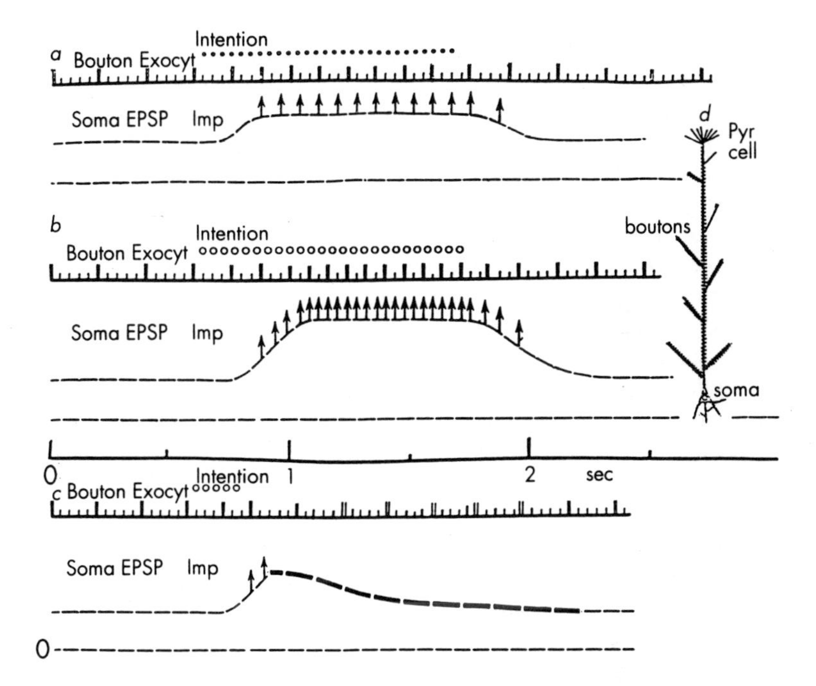

Figure 9.5 Illustration of the proposed mind–brain interaction by intentional increase of the probability for exocytosis. (**a**) The action of a weak intention of about 1 s duration. The *upper part* shows the synaptic input (*short lines*) at a frequency of $50\,s^{-1}$ and the resulting exocytoses (*longer double lines*) for background (1 in 5) and intentionally increased (1 in 3) probabilities. The *lower broken line* is a diagram of the summed milli-EPSPs recorded from the soma. (**b**) The action of a strong intention of the same duration as in (a). The intentionally increased probability here is 1 in 2, resulting in a stronger soma impulse. (**c**) As in (b), but for a brief intention of about 0.2 s. (**d**) A pyramidal lamina V cell with apical dendrite encrusted by the boutons (not all are shown: the actual number is up to 5 000).

Our hypothesis of mind–brain interaction is summarized pictorially in Figure 9.5. Its three components illustrate the effects of a weak (a) and a strong (b) intention (mental acts just over 1 s duration) or of 0.2 s (c). Figure 9.5d illustrates a pyramidal lamina V cell with its apical dendrite encrusted by the boutons of its synaptic input; not all are shown, since they number up to 5 000. In Figure 9.5a–c there are diagrams of the action of a single bouton with the background synaptic input, shown at a frequency of 50 s^{-1} (short lines), that generates exocytoses (shown as longer double lines) to indicate a tubular opening. The initial exocytoses are at 10s^{-1}, giving a probability of 1 in 5 (Redman, personal communication, 1992) for the CA1 hippocampal cells. Below that line is shown digrammatically the EPSP recorded from the soma (d) to the right and produced by electrotonic summation of all the milli-EPSPs on each bouton. In Figure 9.5a the initial EPSP is increased by the action of the mental intention in raising the probability of exocytosis from 1 in 5 to 1 in 3, and that occurs, as has been emphasized earlier, without infringing the conservation laws. This summation is recorded in the soma EPSP and results in the impulse discharges in the soma as shown by the arrows. The 100 or so pyramidal cells of a dendron (Figure 9.2b) would interact in generating the on-going discharges of the SMA neurons (Eccles 1982), and further to the motor pyramidal cells with discharges down the pyramidal tract to the motoneurons (Figure 4.7). The resulting impulse discharges in motor nerves would effect the intended movement with all the control systems cooperating.

In Figure 9.5b there is a stronger intention symbolized by the larger circles and a corresponding increase in probability from 1 in 5 to 1 in 2 with larger EPSPs of the soma and a corresponding increase in the frequency of impulse discharges. In Figure 9.5c the strong intention has a duration of only 0.2 s, yet it generates a brief discharge.

All of Figure 9.5a–c is conventional neuroscience except for the mental intention causing the increased frequency of exocytoses, which is accomplished without violation of the conservation laws. The parameters of Figure 9.5a–c are in accord with conventional neuroscience.

Libet (1990) has studied the time at which a trained subject is intending to move. The conscious will (W) precedes the onset of the movement, the EMG, by about 200 ms. By contrast it has been recognized by Kornhuber, Deecke, and Libet by the averaging technique that a cortical negativity, the readiness potential, precedes the intended movement by as much as 1 000 ms. However, the readiness potential need not indicate a response of some unconscious brain activity preceding the conscious will, that is, that the brain makes the volitional decision, not the mind!

It has been proposed that this problem is illusory because the readiness potential is an *artefact* produced by the averaging technique of recording. It probably arises because the intention to move tends to be on the background of the rising negative phases of the slow EEG waves. Thus the readiness potential results from the cumulative averaging of the background waves of negativity. It can be recognized that it is no more than the tendency of the conscious will to be so located in time in this background. It does not indicate, as has been assumed, that the *brain* initiates the voluntary movement. This is done by the mind in the conscious will (W), which Libet finds to precede the EMG by about 200 ms (Eccles 1990, General Discussion, and Chapter 6) and which is illustrated in Figure 9.5b, c.

This dualistically based diagram (Figure 9.5b, c) illustrates the mind–brain action in a way never attempted by the materialist-monists, who rely on vague statements.

It is important to recognize that, although the intention to move suffers the limitation that it can act only by altering the frequency of exocytoses (Figure 9.5), it can nevertheless control a wide range of movements, both in intensity (Figure 9.5a, b) and in duration (Figure 9.5b, c). More direct actions of the will are precluded by the conservation laws.

Voluntary movement has been explained in principle. This explanation can be extended to the action of all mental influences on the brain, for example in carrying out any planned action, such as speech.

9.6 Conclusions

Based on a careful analysis of neocortical activity we argue that exocytosis is its key mechanism. Exocytosis, the momentary opening of a channel in the presynaptic membrane of a bouton with liberation of the transmitter substance (Figure 9.3f, g), is caused by a nerve impulse. As has been established in many experiments (Redman 1990; Sayer, Friedlander, and Redman 1990; Sayer, Redman, and Andersen 1989), exocytosis is an all-or-nothing-event, occurring with probabilities of the order of 0.25. This observation led us to set up a quantum mechanical model for the trigger mechanism of exocytosis, based on the tunnelling of a quasi-particle representing the trigger.

The quantum treatment of exocytosis links the neocortical activity with the existence of a large number of quantum probability amplitudes, since there are more than 100 000 boutons in the bundle of dendrites known as a dendron (Figure 9.2a, b). In the absence of mental activity these probability amplitudes act independently causing fluctuating EPSPs in the pyramidal cell (Figure 8.6e, g). We put forward the hypothesis that mental intention becomes neurally effective by momentarily increasing the probabilities for exocytoses in a whole dendron (Figure 9.2b) and, in this way, couples the large number of probability amplitudes to produce coherent action.

Our hypothesis offers a natural explanation for voluntary movements caused by mental intentions without violating physical conservation laws. It has been shown experimentally that intentions activate the cerebral cortex in certain well-defined regions prior to the movement (Ingvar 1990; Posner, Petersen, Fox, and Raichle 1988; Roland, Larsen, Lassen, and Skinhøj 1980).

The unitary hypothesis transforms the manner of operation of the intention. One has to recognize that, in a lifetime of learning, the intention to carry out a particular movement would be directed to those particular dendrons of the neocortex that are appropriate for bringing about the required actions. We believe that the proposed hypothesis accounts for action across the mind–brain interface.

Acknowledgements. We acknowledge the help of Dr Stephen Redman in calculations of probabilities of release.

References to Chapter 9

Akert, K., Peper, K., and Sandri, C. (1975) in *Cholinergic Mechanisms*, edited by P. G. Waser (Raven, New York), pp. 43–57.
Deecke, L., and Lang, V. (1990) in *The Principles of Design and Operation of the Brain*, edited by J. C. Eccles and O. Creutzfeld (Experimental Brain Research, Series 21) (Springer, Berlin, Heidelberg), pp. 303–341.
Donald, M. J. (1990) *Proc. Roy. Soc. London A* **424**, 43–93.
Eccles, J. C. (1982) *Arch. Psychiatr. Nervenkrank.* **231**, 423–441.
Eccles, J. C. (1990) *Proc. Roy. Soc. London B* **240**, 433–451.
Edelman, G. M. (1989) *The Remembered Present: A Biological Theory of Consciousness* (Basic Books, New York).
Einstein, A., Rosen, N., and Podolsky, B. (1935) *Phys. Rev.* **47**, 777–780.
Everett, H. (1957) *Rev. Mod. Phys.* **29**, 454–462.
Gray, E. G. (1982) *Trends Neurosci.* **5**, 5–6.
Ingvar, D. H. (1990) in *The Principles of Design and Operation of the Brain*, edited by J. C. Eccles and O. Creutzfeld (Experimental Brain Research, Series 21) Springer, Berlin, Heidelberg), pp. 433–453.
Jack, J. J. B., Redman, S. J., and Wong, K. (1981) *J. Physiol. London* **321**, 65–96.
Jessel, T. M., Kandel, E. R., Lewin, B., and Reid, L. (eds) (1993) Signaling at the synapse, Review Supplement to *Cell* **72**/*Neuron* **10**.
Kelly, R. B., Deutsch, J. W., Carlson, S. S., and Wagner, J. A. (1979) *Ann. Rev. Neurosci.* **2**, 399–446.
Korn, H., and Faber, D. S. (1987) in *New Insights into Synaptic Function*, edited by G. M. Edelman, W. E. Gall, and W. M. Cowan (Wiley, New York), pp. 57–108.
Libet, B. (1990) in *The Principles of Design and Operation of the Brain*, edited by J. C. Eccles and O. Creutzfeld (Experimental Brain Research, Series 21) (Springer, Berlin, Heidelberg), pp. 185–205, plus General Discussion, pp. 207–211.
Margenau, H. (1984) *The Miracle of Existence* (Ox Bow, Woodbridge, CT).
Messiah, A. (1961) *Quantum Mechanics* (North Holland, Amsterdam), pp. 231–242.
Mountcastle, V. B. (1978) in *The Mindful Brain*, edited by F. C. Schmitt (MIT Press, Cambridge MA), pp. 7–50.

Peters, A., and Kara, D. A. (1987) *J. Comp. Neurol.* **260**, 573–590.
Pfenninger, K., Sandri, C., Akert, K., and Eugster, C. H. (1969) *Brain Res.* **12**, 10–18.
Popper, K. (1990) *A World of Propensities* (Thoemmes, Bristol).
Popper, K., and Eccles, J. C. (1977) *The Self and Its Brain* (Springer, Berlin, Heidelberg).
Posner, M. I., Petersen, S. E., Fox, P. T., and Raichle, M. E. (1988) *Science* **240**, 1627–1631.
Redman, S. J. (1990) *Physiolog. Rev.* **70**, 165–198.
Roland, P. E., Larsen, B., Lassen, N. A., and Skinhøj, E. (1980) *J. Neurophysiol.* **43**, 118–136.
Sayer, R. J., Friedlander, M. J., and Redman, S. J. (1990) *J. Neurosci.* **10**, 626–636.
Sayer, R. J., Redman, S. J., and Andersen, P. (1989) *J. Neurosci.* **9**, 845–850.
Schmolke, C., and Fleischhauer, K. (1984) *Anat. Embryol.* **169**, 125–132.
Squires, E. J. (1988) *Found. Phys. Lett.* **1**, 13–20.
Stapp, H. P. (1991) *Found. Phys.* **21**, 1451–1477.
Szentágothai, J. (1978) *Proc. Roy. Soc. London B* **201**, 219–248.
von Neumann, J. (1955) *Mathematical Foundations of Quantum Mechanics* (Princeton University Press, Princeton NJ), Chap. IV.
Walmsley, B., Edwards, F. R., and Tracey, D. J. (1987) *J. Neurosci.* **7**, 1037–1046.
Wigner, E. P. (1967) *Symmetries and Reflections* (Indiana University Press, Bloomington, IN), pp. 153–184.

10 The Self and Its Brain: The Ultimate Synthesis

10.1 Life and Mind

There were two absolutely unpredictable happenings in the story of the cosmos. The first was the origin of life, the second the origin of mind. If one asks where was mind or consciousness before it came to be experienced with the mammalian brain—about 200 million years ago (Chapter 7)—the answer would be the same as to the question where was life before it came to exist on planet Earth about 3.4 billion years ago. Both of these origins led to transcendent developments: for the one the whole biological world of evolution culminating in *Homo sapiens sapiens* about 90 million years ago (Eccles 1989); for the other the *conscious world* of mammals (Chapter 7) leading on to hominid evolution (Eccles 1989, 1992 and Chapter 7) and so to the great wonders of human culture, which is the greatly embellished World 3 of Popper. This is the world of selves, each unique self with its own unique human brain.

10.2 Self–Brain Dualism

Self–brain dualism demands primarily two authentic orders of existents with completely independent ontologies. As has been long recognized (Popper and Eccles 1977), this dualism necessitates transactions across the mind–brain interface in both directions (Figures 1.3, 5.5, and 6.1). Such formidable problems have led to various types of evasions, which are exemplified in Figure 1.2 by the four materialist theories of the mind. The only one of present interest is the identity theory (Feigl 1967)(see Chapter 1.3) in which the brain–mind problem is resolved by the proposed 'identity' of mental events with neural events in the activities of the higher centres of the brain, which is essentially a materialist hypothesis.

This strange postulate of identity is never explained, but it is believed that it will be resolved when we have more complete scientific understanding of the brain, perhaps in hundreds of years; hence this belief is ironically termed 'promissory materialism' (Popper and Eccles 1977). In Chapter 3 most of those discussed were identity theorists: Changeux, Crick and Koch, Dennett, Edelman, Sperry, and Searle.

At the end of Chapter 2 I described how, after decades dedicated to dualism and with a progressively increasing understanding of the neocortex, I still felt blocked by the conservation laws that proscribed any action on the brain by non-material mental events. Margenau's book *The Miracle of Existence* was a light at the end of the tunnel with his suggestion that quantum physics could provide an understanding of how the mind could interact with the brain without requiring energy. I was overwhelmed with this message and proceeded to show how the most subtle understanding of the structure and function of the microsites of the neocortex could lead to important insights into the mind–brain interaction.

In Chapters 4 and 5 self–brain dualism again seemed to be a tenable hypothesis; but it is the dualism of Worlds 1 and 2 (Popper and Eccles 1977) and *not* the Cartesian dualism of Dennett and others, for example, in Chapter 3.

10.3 The Self

As I develop and illustrate the dualist philosophy there is concentration on the self, which is the unique experience of each of us throughout our whole lifetime. This book was introduced with Sherrington's poetic description of the self in Chapter 1.

We now have convincing scientific demonstrations of how the self with pure ideation (Ingvar 1989) is effective in activating selected regions of the cerebral cortex (Figures 5.2, 5.4, 10.1, 10.2, and 10.3). So lavish is this mental control of cerebral activity (see Chapter 10.5 on attention) that we can assume a complete dominance of the brain by the self. And now the hypothesis has been proposed, for the first time, *how* these mental influences could control brain activities without infringing the conservation laws (Beck and Eccles 1992 and Chapter 9 and Figures 9.5 and 10.2). So the materialist criticism of dualism by Dennett, Changeux, and Edelman loses its scientific basis. Materialist explanation of the mind–brain problem, such as the identity theory, can now be recognized as unscientific and even as superstitions that have lingered on too long, as is also the case for promissory materialism. All now seem to be untenable.

Each of us has naturally the dualist belief of self and brain interaction, the so-called folk philosophy or psychology, but rejection has been demanded

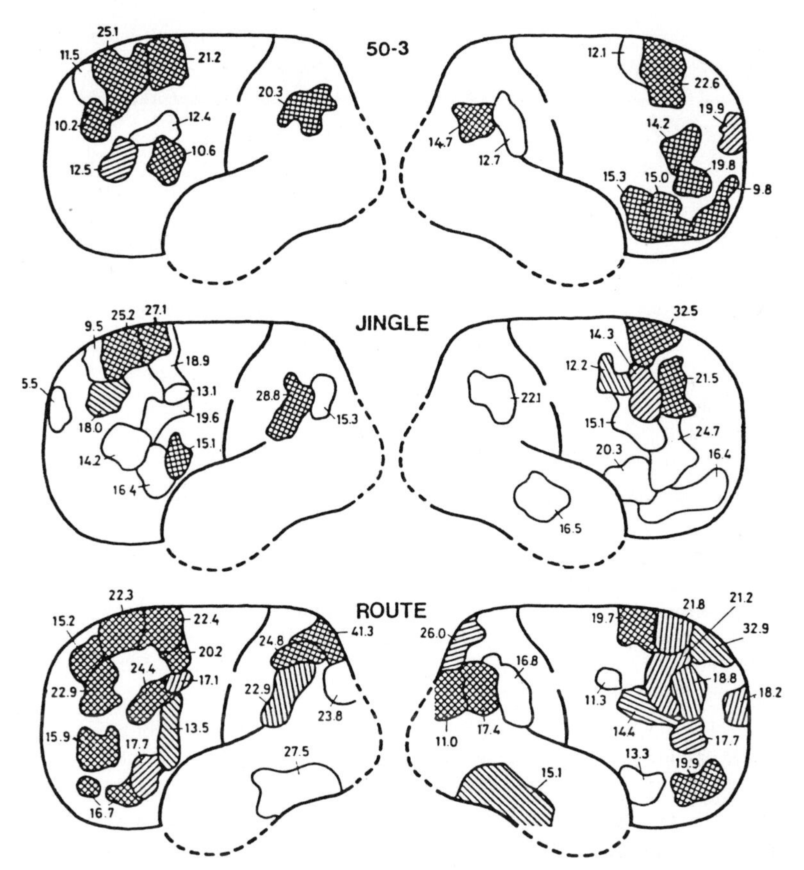

Figure 10.1 Mean increases of rCBF as a percentage and their average distribution in the cerebral cortex under three different conditions of silent thinking, as described in the text. Left hemisphere: six subjects; right hemisphere: five subjects. *Cross-hatched areas* have rCBF increases significant at the 0.005 level. With *hatched areas* $P < 0.01$ and with *outlined areas* $P < 0.5$) (Roland and Friberg 1985).

by the prevailing materialist reductionist philosophy. This philosophy is espoused in various guises by neuroscientists without being subjected to the philosophical examination that was done by Popper and myself (1977). It is a naive philosophical belief, as Edelman (1989, p. 12) seems to admit. But it achieved the status of a materialist article of faith, because it claimed to account for all experience on the basis of various versions of materialism, as can be seen in Chapter 3. All of this pseudophilosophy can now be rejected in the light of the hypothesis that the self–brain interaction occurs

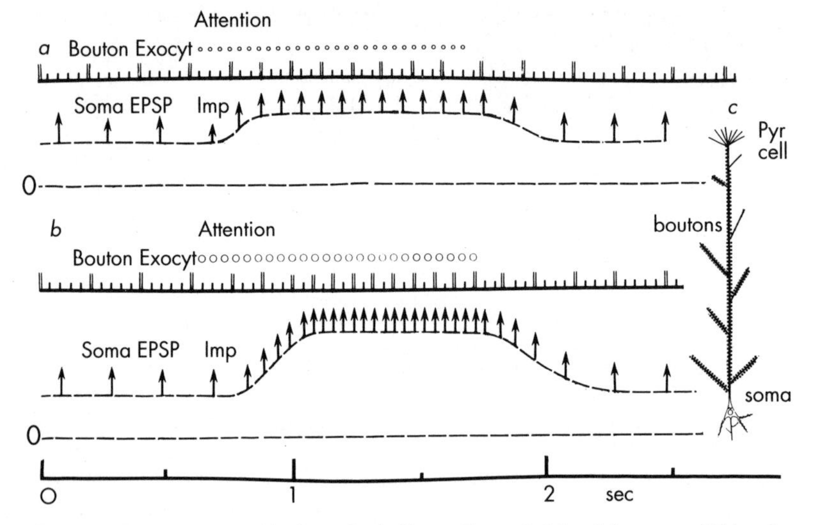

Figure 10.2 In general, this figure is similar to Figure 9.5, but it has two additional features. (1) As indicated, the mental influence shown by a *row of circles* of about 1 s duration is an attention. (2) The background EPSP is shown to induce in the soma a background low-frequency-impulse discharge. If the probability were 1.0, there would be a *background exocytosis* for every impulse in the *upper row*. So the mental attention could have no action.

without infringing the conservation laws of physics (Beck and Eccles 1992 and Chapter 9). So Searle (1984, p. 99) (see Chapters 3.8 and 3.9) can continue with a good conscience to engage in voluntary, free, intentional actions!

The self action on the brain (neocortex) can presumably be extended by suitable experiments using rCBF and PET scanning (see Chapter 10.5 below and Figure 10.3) to include all of our conscious experiences, even the most subtle and transcendent. It is reassuring that all of the richness and enjoyment of our experiences can now be accepted without any qualms of conscience that we may be infringing conservation laws!

It is strange to find how materialists such as Changeux, Edelman, and Dennett (Chapter 3) engage in the most convoluted linguistics to derive spiritual values and ideals from their barren materialist beliefs.

The self–brain interaction opens up the most wonderful future for the understanding of the cerebral cortex. Psychon–dendron interactions provide the basic happenings. The low levels of exocytotic probability for cortical neurons in the three reliable hippocampal experiments (Figure 8.6) (Sayer et al. 1989, 1990) provide a valuable base for the proposed increase of

EPSPs (Beck and Eccles 1992 and Chapter 9) by the whole range of mental influences (Figures 9.5 and 10.2). It will be recognized that I am not giving the brain any mental properties or qualia. These are exclusive to the self as being the Worlds 2 and 3 of Figures 1.1, 1.3, and 6.1.

In Chapter 3.9, I quoted Searle (1992, p. 55):

> The deepest motive for materialism . . . is simply a terror of consciousness.

Searle thinks that this arises because of the feature of subjectivity that threatens the objectivity as espoused by Monod (1971):

> How can we possibly give a coherent world picture if the world contains these mysterious conscious entities. Yet we all know that we are for most of our lives conscious, and that people around us are conscious[!] (p. 95)

But I think that there is a more personal reason for this terror that derives from the uniqueness of the experiencing self with its religious overtones as developed with the theme of freedom of the will and moral responsibility. Reference should be made to the valuable extracts from Hodgson (1992, Chapter 19) in Chapter 3.6.

By contrast, materialists believe in the identity theory operating with neural networks so that consciousness can be downgraded to a materialist performance. So contemporary philosophy and neurophilosophy become a materialist cacophony of structuralism and functionalism culminating in roboticism! The theme song should be *The Sorcerer's Apprentice*!

10.4 Ideation and Free Will

Ingvar (1989) has given much further evidence of the activation of the human brain during pure sensory ideation, as in Figure 10.1, and pure motor ideation, as in Figure 5.2b.

This extraordinary effectiveness of self in activating the neocortex can now be explained by the increases induced in probabilities of exocytoses in the presynaptic vesicular grids (PVGs) of the hundreds of thousands (Figures 7.2, 7.3, and 8.3) of boutons on one dendron. According to the psychon hypothesis, each psychon has available in its linked dendron (Figure 6.10) an enormous number of PVGs with their quantal probabilities for exocytoses.

The widespread influence of the mind on the brain is well illustrated in Figure 10.1 (Roland and Friberg 1985), where there was silent thinking of three distinct types with recording of the increased cerebral circulation by the radio-xenon procedures.

In Figure 10.1a for the 50 – 3 frame the subject was silently carrying out sequential subtractions of 3s from 50, the memory procedure being 50 . . . 47 . . . 44 . . . 41, etc., continuing on below 2 to –1, –4, etc., until after the end of the recording of the regional cerebral blood flow (rCBF) at about 45 s. During all that period the subject was undisturbed by the environment. It is remarkable that the significantly increased rCBFs were on both sides in the prefrontal lobe, except for the angular gyrus in the parietal lobe with increases of 20.3 per cent and 14.7 per cent. Correspondingly the clinical condition acalculia has been observed with bilateral destruction of the angular gyrus.

In Figure 10.1b, the 'jingle' frame shows that the rCBF increases when the silent memory was concentrated on a task of jumping mentally to every second word of a well-known Danish nonsense-word sequence or jingle, which consists of a closed loop of nine words. Again, almost all of the activated cortical areas were in the frontal lobe.

In Figure 10.1c for the 'route' frame the subjects had the task of visualizing the successive scenes as they travelled in their imagination along a well-known street route.

However, in interpreting Figure 10.1c there is no need to consider all of this activity as being *primarily* induced by the task. There is close neural linkage of many of these prefrontal areas, and primary activation of some of the mental–visual task would result in secondary or tertiary activation of the remainder. It is important to recognize the wide range of the prefrontal excitation involved in some visual memory task.

The demonstration of the extraordinary effectiveness of thinking on the cerebral cortex (Figure 10.1) leads to the question: how was this enormous influence of the self on the brain created? The answer is: it has been a lifetime of active learning. The earliest stage is exhibited by a baby in its cot viewing its hands. It is learning to move its hand as it wills with progressively more skill in touching and grasping. The mind of the baby brings about intended movements with motor cortical control as in Figure 5.2. Throughout life all skills are learnt in this way by devoted attention.

There has long been a question about voluntary movement: materialists have asserted that a *mental intention* cannot bring about a voluntary movement because that is contrary to the conservation laws. However, Searle (1986) writes well on free-will and correctly believes that it is tied to consciousness despite the conservation laws. But now with the work of Beck and myself (1992 and Chapter 9) there is no such proscription. In fact we are presented with the wonderful exuberance of relating to our brains with full freedom, for action and imagination. All normal human selves behave with that freedom despite the materialist's and the monist's denial of this freedom. It is the age-old problem of the freedom of the will. We can now transcend that problem with the hypothesis formulated in Chapter 9 that the self can act effectively on the brain without infringing the conservation laws. We can rejoice in our new-found freedom, which is based on the dualism of Figures 1.1, 1.3, and 6.1. In Figures 9.5 and 10.2 there is interaction of two distinct entities, the spiritual self (World 2) and the material brain (World 1), as defined by Popper and Eccles (1977).

10.5 Attention

The rCBF investigation by Roland (1981) on the effect of mental attention on the neocortex has already been illustrated (Figure 6.12). Attention was focused on a finger tip in expectation of a just detectable touch. The increased cortical activity appropriately was largely on the finger-touch area (Figure 5.4a). If the expectation was on the lips, then the increased cortical action was largely on the lips' somatosensory area.

The positron emission tomography (PET) studies by Raichle and his associates (Posner et al. 1985; Corbetta et al. 1990; Pardo et al. 1991) (Figure 10.3) provide most important evidence on the manner in which attention can influence the neocortex of conscious human subjects. There was a special attempt to localize the cerebral areas activated in cognition. For example,

> visual imagery, word reading and even shifting visual attention from one location to another are not performed by any single brain area. Each of them involves a large number of component computations that must be orchestrated to perform the cognitive task! (Posner et al. 1985, p. 1630)

However, selective attention appears to use neural systems separate from those involved in passively collecting information about a stimulus.

> If active selection or visual search is required, this is done by a spatial system that is deficient in patients with lesions of the parietal lobe. Similarly, in the anterior brain, the lateral left frontal lobe is involved in the semantic network for coding word associations. Local areas within the anterior cingulate become increasingly involved when the output of the computations within the semantic network is to be selected as a relevant target. Thus the anterior cingulate is involved in the computations in selecting language or other forms of information for 'action'.
>
> In the study of visual imagery, models distinguish between a set of operations involved in the generation of an image and those involved in scanning the image once it is generated. (Posner et al. 1985, p. 1630)

By PET scanning, Corbetta et al. (1990) showed that

> attention enhanced the activity of different regions of extrastriate visual cortex that appear to be specialized for processing information related to the selected attribute. (p. 1556)

For example, attention to a red hat in a crowd improves accuracy and sensitivity on visual detection or discrimination tasks.

> Attending to a visual attribute, such as its color, might be expected to modulate neuronal activity in brain areas that are specialized for processing that attribute, such as V_4 for color (Zeki 1976). Different regions or the extrastriate visual cortex were activated when attending to different attributes of a visual display. (Corbetta et al. 1990, p. 1556)

> Attention to basic visual attributes, such as shape, color, or velocity, appears to influence behavioral and physiological measures of visual processing. Behaviorally, sensitivity for making subtle discriminations is increased by attention. Physiologically, neural activity is increased in extrastriate regions specialized for processing information related to the selected visual attribute. These enhancements reflect cognitive (top-down) control of visual processing. (Corbetta et al. 1990, p. 1558)

Here is a clear affirmation of the concept that mental events (attention) can activate neural events as is indicated in Figure 10.2, which is a slight modification of Figure 9.5.

The four beautiful colour displays of Figure 10.3 give wonderful illustrations of the specific cerebral activities induced by a wide variety of attentions and selections during linguistic performances. It can be predicted that eventually all of the neocortex will be shown to respond focally by specific activities, by using this PET scanning technique.

The limited testings so far studied have revealed a wide range of cortical activation, so it can be expected that *all parts of the cortex eventually will be found to be activated by appropriate attentions.*

Figure 10.2 is an explanatory diagram to show attention acting similarly to intention in Figure 9.5, with the additional feature that the background exocytoses not only generate a sustained EPSP by summation of the milli-EPSPs but that this background EPSP generates a background impulse discharge. The attention of about 1 s duration acts like the intention of Figure 9.5 to generate a burst of impulse discharges, which is especially intense for the stronger attention in Figure 10.2b. This diagram incorporates the hypotheses of Figure 9.5 with attention effectively increasing the bouton exocytosis without infringing the conservation laws (Beck and Eccles 1992 and Chapter 9).

It can be hypothesized that a self is able by attention to activate any selected parts of the neocortex at will. This selective attentional activation

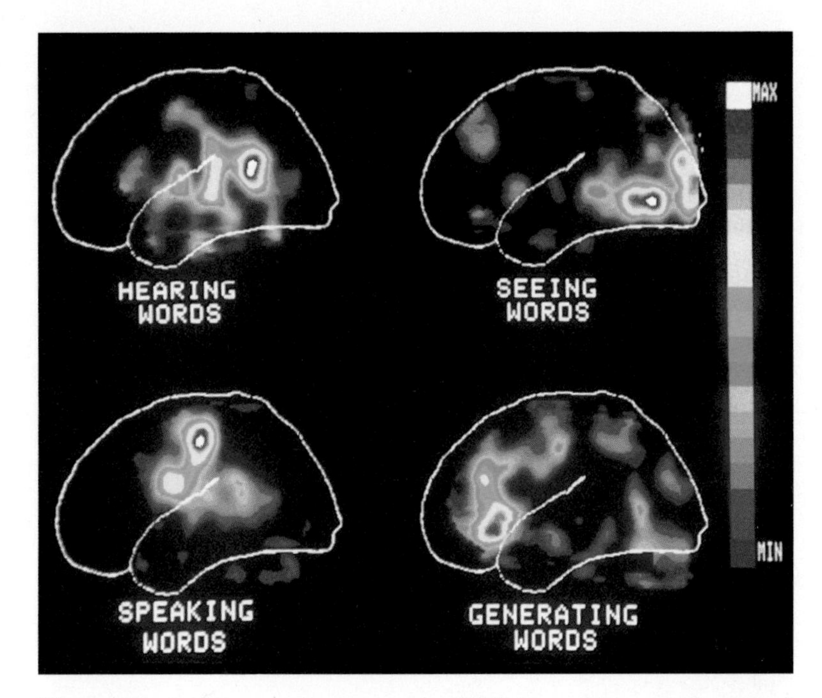

Figure 10.3 Positron emission tomography (PET) scans of the cerebral cortex of conscious human subjects performing four intellectual tasks relating to words. The PET technique reveals that the blood flow signalling the activity of the neocortex is sharply focused in specific locations depending on the linguistic tasks being performed. The *colour scale to the right* shows the wide range of cortical activities. (Dr Marcus Raichle of the Washington University School of Medicine in St Louis made the images in 1992 and has kindly provided them for my book.)

is of vital importance in the whole range of our experiences. To take a trivial example, we notice by chance a bee landing on a flower. If we wish to observe more closely its action on the flower, our attention by psychon activation will enhance the range and intensity of the dendron activation for increasing the visual input and discrimination. It can be recognized that by attention we enhance and embellish our perceptions. Throughout our lifetime each of us has learned to use our brain with finesse and skill. This can be appreciated when we listen attentively to music or examine a beautiful picture or enjoy some beauty in nature. Attention is specially related to the appreciation of beauty. This transcendent experience evoked from the brain by attention is the basis of our character and personality. Our self learned precisely how to 'play' with any chosen part of our neocortex. It is a psychon–dendron game. Of course we have no knowledge of the anatomy of this self–brain dialogue. It is a functional performance with assurance derived from our life-long experiences. We can recognize that the action of self on brain is illustrated for three very different thought processes in Figure 10.1 (Roland and Friberg 1985) and particularly in Figure 10.3.

Dennett and Kinsbourne (1992) considered in great detail the problem: where in the brain is consciousness experienced? Their comprehensive introduction was followed by 22 replies from chosen experts. Dennett and Kinsbourne elaborated Dennett's Multiple Drafts Hypothesis (see Chapter 3.4 on Dennett) as against the alternative unification in the so-called Cartesian Theater. Other alternative locations were suggested.

In the light of the cortical performance in attention and thinking, an unexpected answer can be given: *consciousness is experienced in the brain where you evoke it by your attention*, which plays on selected areas of the cerebral cortex to give excitation. That excitation leads to amplified dendron responses to sensory inputs and so to psychon activations and consciousness. Superimposed on this simple attentional operation there would be a continuing dialogue between attention by the self and the selected neocortical areas with their sensory inputs. Dennett and Kinsbourne may feel able to accept these localizing processes. It would account for the diversity of the Multiple Drafts locations.

10.6 Artificial Intelligence—AI

Tragically and absurdly misunderstood is the wonder and mystery of the human self with its spiritual values, with its creativity, and with its uniqueness for each of us.

Penrose in his book *The Emperor's New Mind*, which I have abstracted from in Chapter 3.7, has exposed very well the basic errors in the AI story. Of course I am not condemning the use of computers even of the most sophisticated kind. They are essential tools in our efforts to understand more and more the wonderful workings of the human brain. I particularly want to criticize the erroneous claims of the robotics advocates that robotics will lead to the understanding and simulation of human selves. It is all very clever, as, for example, Edelman's Darwin 3, but it is for display and grant support! It is of no value in helping to understand the brain. It is very revealing to discover in all the massive literature on consciousness and AI that there is no reference to the three fundamental discoveries that are the basis of the scientific hypothesis of the self–brain interaction in Chapter 9 (see Figures 9.5 and 10.2):

1 *The presynaptic vesicular grid with its exquisite design for exocytotic probability.*
2 *The exocytotic probability with its low level that gives great opportunity for enhancement by the mind without infringing the conservation laws.*
3 *The bundling of apical dendrites of pyramidal cells into dendrons that provide the necessary great amplification of the milli-EPSPs produced by each exocytosis.*

It is amazing and very disturbing that all the great scientific efforts in the attempt to understand the brain display blindness to the microstructures and microfunctions of the neocortex. Penrose (1991) did realize the necessity for scientific study of the neocortex, but he did not discover the essential microlevels of understanding. Sperry was hovering in the right direction in his rejection of materialism, but he missed the necessary microunderstanding.

Searle (1984) rejects the programmes and claims of the artificial intelligence advocates. As Searle states:

> The programming of a computer is entirely syntactical, but the mind has more than a syntax, it has a semantics. (p. 31)

Searle concludes:

> No computer program by itself is sufficient to give a system a mind. Programs, in short, are not minds, and they are not by themselves sufficient for having minds. The project of trying to create minds solely by designing programs is doomed from the start; whereas consciousness, thought, feelings, emotions and all the rest of it involve more than a syntax. These features, by definition, the computer is unable to *duplicate*, however powerful may be its ability to *simulate*. And no simulation by itself ever constitutes duplication. (p. 37)

As stated in Chapter 9 all the quantum physicists failed because their approach got no further than generalities, rather like electricity before electrons and light before photons. This scientific failure can be attributed to the prevailing materialism with its superstitious beliefs.

What I find particularly disturbing is the claim of the AI operators that they are on the verge of supercomputers that will experience consciousness. Many years ago at a Yale University conference I asked Minsky of M.I.T., the most eloquent of the hard-AI advocates, why they made claims to have supercomputers that would be conscious. His revealing reply was:

> Because I get larger grant support!

10.7 The Psychon World

The most extraordinary experience for neuroscientists is the unification of the most diverse experiences as they are apprehended from moment to moment. The unification of thought was recognized by the great psychologist William James over 100 years go. As Stapp states (see Chapter 3.12):

> James's principal claim at the fundamental level is the wholeness or unity of each conscious thought.

James made an exhaustive study by introspection. He recognized that the physics and brain sciences of his day were too primitive to account for the consciousness associated with the brain. Now over a hundred years later we can respond to James's request!

Vision provides the most extraordinary of our perceptual experiences. The inverted image on the retina is subjected to a multitude of analytical procedures in sequential and parallel performance of the visual cortex, where such features as slope, direction, movement, form, contrast, intensity, and colour are selected for analysis, but nowhere in the brain is there synthesis to the original retinal image, except for an infrequent crude response for face or hands in some neurons of the inferior temporal lobe. Yet the original image is experienced in the mind with stereopsis. There is now an attempt to discover neurons of the visual cortices with some organization of association (Singer 1990). It is called the binding problem, but it is no more than a beginning. It has to be recognized that the perfect experiences of visual images are in the mind that seems to synthesize from the analysis in the visual cortex. Possibly the dendron–psychon linkage may give an insight into this extraordinary learned phenomenon. It can be conjectured that intense learning procedures are involved in giving the perfection of the visual experience that we recognize from moment to moment as qualia. Doubtless the perfection of auditory or tactual or movement experiences have similar explanations, as considered towards the end of Chapter 6.

This extraordinary experience of a global or landscape vision would be of great evolutionary advantage to mammals over other animals that had a limited discoordinated operation of the visual inputs.

Psychons are units of experience and so have a dominant reality. The evolution of consciousness was of great evolutionary advantage, as described in Chapter 7.

It can be accepted that all mammals are conscious beings with some conscious control of their actions and some conscious experiences (Eccles 1989, p. 173). The dendron–psychon interaction is thus essential to their mental life. The human situation is a further development with the coming of self-consciousness in which it can be conjectured that psychons may exist apart from dendrons in a unique psychon world, which is the world of the self and which is shown centrally in Figure 6.1.

There are great unknowns in this conjectured world of psychons. Their very nature is to be experiences, and we can only indicate their existence diagrammatically by the site of their dendron interaction, which is shown as an ensheathing in Figure 6.10 for three types of psychon.

Transmission of psychon to psychon could explain the unity of our perception and of the inner world of our mind that we continually experience from moment to moment, as was recognized by William James. This unity can be experienced for all the happenings of World 2.

Hitherto it has been beyond explanation by any mind–brain theory that multifarious neural events in our cerebral cortex can from moment to moment give us global mental experiences that have a unitary character. We feel central to our experiential world (World 2). This phenomenon is in the central core of World 2 in Figure 6.1 with the labelling: psyche, self, soul. Arrows are shown projecting into this central locus from the region of outer sense and inner sense and the dendrons of the neocortex. This raises a fundamental question: are the experiences of the self also composed of unitary psychons in the same manner as for perceptual and other experiences? If so, is each of these psychons also linked with its dendron, and where in the neocortex are these dendrons? We can further ask if there is a category of organized psychons not linked to dendrons, but only with other psychons forming a psychon world apart from the brain.

It has earlier (Chapter 10.5) been described how the self can act on and influence the happenings in the neocortex with subtlety and understanding. It seems that the self must be cognizant of all the learnt potentialities of the brain. It must have the memory to act with speed and skill in calling into action the appropriate stored memories in the brain. So the psychon complex of the self (Figure 6.1) must be endowed also with 'memories' that are continuously up-dated for efficient action. Here is a fundamental psychological problem for investigation.

10.8 The Meaning of Conscious Existence as a Self

Problems relating to the experienced uniqueness of each self are neglected in contemporary philosophy. Presumably this arises from the pervasive materialism, which is blind to the fundamental problems arising in spiritual experience, as emphasized by Searle (see Chapters 3.8 and 3.9). I very much approve of the quotation from Hodgson on the search for purpose in life and its religious meaning (see Chapter 3.6).

The most usual materialist statement is that the experienced uniqueness derives from the genetic uniqueness. The unique genome that is alleged to be the basis of the experienced uniqueness is the consequence of an infinitely improbable genetic lottery (even $10^{15\,000}$ against, on the conservative estimate of 50 000 human genes (Eccles 1989, p. 236)). Moreover, as Stent (1981) has pointed out, the phenotypic development of the brain is far removed from the genotypic instructions because of the operations of what Waddington (1969) has termed 'development noise'. For example, the genotype is involved in the building of the brain, but it acts in an environment that profoundly modulated its phenotypic building process. With identical twins the identical genomes would contribute to the building of different brains because of the diversity of developmental noise.

A frequent and superficially plausible answer to this enigma is the assertion that the determining factor is the uniqueness of the accumulated experiences of a self throughout its lifetime. It is readily agreed that our behaviour and memories and in fact the whole content of our inner conscious life are dependent on the accumulated experiences of our lives; but no matter how extreme the change, at some particular decision point, which can be produced by the exigencies of circumstances, one would still be the same self able to trace back one's continuity in memory to the earliest remembrances at the age of one year or so, the same self in quite another guise. There could be no elimination of a self and creation of a new self!

Since materialist solutions fail to account for our experienced uniqueness, I am constrained to attribute the uniqueness of the self or soul to a supernatural spiritual creation (Eccles 1989, p. 238).

It is the certainty of the inner core of unique individuality that necessitates the 'divine creation'. I submit that no other explanation is tenable; neither genetic uniqueness, with its fantastically impossible lottery, nor environment differentiations, which do not *determine* one's uniqueness but merely modify it.

This conclusion is of inestimable theological significance. It strongly reinforces our belief in the human soul and in its miraculous origin in a divine creation. There is recognition not only of the transcendent God, the creator of the cosmos, the God in which Einstein believed, but also of the immanent God to whom we owe our being.

I here express my efforts to understand with deep humility a self, myself, as an experiencing being. I offer it in the hope that we human selves may discover a transforming faith in the meaning and significance of this wonderful adventure that each of us is given on this salubrious Earth of ours, each with our wonderful brain, which is ours to control and use for our memory and enjoyment and creativity and with love for other human selves.

As Pascal so wonderfully opined, each of us comes to exist as a self, at a time and place, beyond our comprehension. Why here and not elsewhere? Why now and not another time? Are we not participants in the meaning, where otherwise there is no meaning? Do we not experience and delight in fellowship, joy, harmony, truth, love, and beauty, where there is otherwise only the mindless universe?

10.9 *How*

The title of this book *How the Self Controls Its Brain* was chosen to be a clear challenge. It also harks back to the book I wrote with Popper: *The Self and Its Brain* (1977).

It contrasts with the convoluted generalities of attempted materialist solutions for the origin of consciousness, as is illustrated in the abstracts in Chapter 3 for Changeux, Dennett, and Edelman, after explanations are given with their admission that the *How* is unknown.

The strategy of the materialists, both neuroscientists and philosophers, is either to ignore the philosophy of dualism that Popper and I developed in our 1977 book or to confuse our dualism with the ancient two-substance dualism of Descartes. Furthermore, they reject the dualistic philosophy that I have recently published (Eccles 1989), on the grounds that it infringes the conservation laws of physics.

It is now shown by Beck and myself (1992 and Chapter 9) that the dualism I present is based scientifically on quantum physics and is in accord with the conservation laws. Will the materialists dare to reject it? As Searle so wisely states, materialists have a terror of consciousness. If so, their philosophy becomes emotional and is not rational. It is irrational. Even the famed objectivity of Monod's dogmatism turns out to be emotional in the last chapter of his 1971 book, called appropriately *The Kingdom or the Darkness*. Monod has given over his rationality for his ultimate role of a prophet, as has been recognized in several reviews of his book by Thorpe, Cournand, and others.

Now *for the first time* in the article by Beck and myself (1992) that forms Chapter 9 there is presented the scientific hypothesis of *how* the self may control its brain without infringing the conservation laws. This is our continual dualist experience in all our waking hours. It contradicts the repeated statement by Searle (1984) (see Chapter 3.8) that

Brains cause Minds,

which is the materialist identity theory. So I present this book as a challenge to all materialists.

This dualist conclusion is the culmination of my life-long dedication to the self–brain problem, which is described in the present comprehensive book with a historical appraisal. So I dare to have 'How' in the title!

There is an extraordinary consequence of the hypothesis that mental events (psychons) effectively act on the dendrons by increasing the probability of the exocytoses generated by invasion of a bouton by presynaptic impulses. Fortunately the quantal probability is low (0.3–0.4, Chapter 8.6) for the cerebral cortex, the hippocampus. If the probability were as high as 1.0, mental experiences could have no effective action on neural events of the dendrons. The evolutionary development of the mammalian neocortex would not have redeemed the mindlessness of the brain, as described in Chapter 7. There would be no 'HOW' of conscious experiences coming to exist as an eventual outcome of the biological evolution of the mammalian brain. All depends on the neural design of the ultramicrosite operation with the low exocytotic probability of the millions of boutons in the mammalian neocortex and in the coming-to-be of some primitive conscious experiences that achieve expression because of the low exocytotic probability.

Human self-consciousness depends on the miraculous coming-to-be of the Self (Chapter 10.8) that achieves expression by enhancing the low exocytotic probability of the billions of boutons of the human neocortex.

Our hypothesis thus gives the ultimate 'HOW' of the human neocortex and so redeems what would otherwise be a mindless world peopled by unconscious beings.

A more general hypothesis is that so long as transmission between neurons was electrical, it was explicable by classical physics and so would provide no basis for the influence of a 'latent mental world', from Stapp's wise statements, as given in Chapter 3.12. The mental world could play a key role only when chemical transmission had evolved with chemical transmitters packaged in synaptic vesicles and with conservation by controlled exocytosis in which quantum physics could participate. This mysterious evolutionary process gave the opportunity for the highly evolved cerebral cortex of mammals to be open to mental influences giving consciousness and so illuminating the hitherto mindless world. So chemical synaptic transmission is the fundamental basis for our conscious world and all its transcendent creativity, which we value in the conscious self with its control of the brain in the 'HOW' of this book.

References to Chapter 10

Beck, F., and Eccles, J. C. (1992) Quantum aspects of brain activity and the role of consciousness, *Proc. Nat. Acad. Sci.* **89**, 11357.

Corbetta, M., Miezin, F. M., Dobmeyer, S., Shulman, G. L., and Petersen, S. E. (1990) Attentional modulation of neural processing of shape, color and velocity in humans, *Science* **248**, 1356–1359.

Dennett, D., and Kinsbourne, M. (1992) Time and the observer. The where and when of consciousness in the brain, *Behavioral and Brain Sci.* **15**, 185–247.

Eccles, J. C. (1989) *Evolution of the Brain: Creation of the Self* (Routledge, London).

Edelman, G. M. (1989) *The Remembered Present: A Biological Theory of Consciousness* (Basic Books, New York).

Feigl, H. (1967) *The 'Mental' and the 'Physical'* (University of Minnesota Press, Minneapolis MN).

Hodgson, D. (1991) *The Mind Matters* (Clarendon, Oxford).

Ingvar, D. H. (1990) On ideation and 'ideography', in *The Principles of Design and Operation of the Brain*, edited by J. C. Eccles and O. Creutzfeldt (*Experimental Brain Research*, Series 21) (Springer, Berlin, Heidelberg), pp. 433–453.

Margenau, H. (1984) *The Miracle of Existence* (Ox Bow, Woodbridge CT).

Monod, J. (1971) *Chance and Necessity* (Knoff, New York).

Pardo, J. V., Fox, P. T., and Raichle, M. E. (1991) Localization of a human system for sustained attention by positron emission tomography, *Nature* **349**, 61–64.

Penrose, R. (1989) *The Emperor's New Mind: Concerning Computers, Minds, and the Laws of Physics* (Oxford University Press, Oxford).

Popper, K. R., and Eccles, J. C. (1977) *The Self and Its Brain* (Springer, Berlin, Heidelberg).

Posner, M. I., Petersen, S. E., Fox, P. T., and Raichle, M. E. (1985) Localization of cognitive operations in the human brain, *Science* **240**, 1627–1631.

Roland, P. E. (1981) Somatotopical tuning of postcentral gyrus during focal attention in man. A regional cerebral blood flow study, *J. Neurophysiol.* **46**, 744–754.

Roland, P. E., and Friberg, L. (1985) Localization in cortical areas activated by thinking, *J. Neurophysiol.* **53**, 1219–1243.

Searle, J. R. (1984) *Minds, Brains and Science* (British Broadcasting Corporation, London).

Searle, J. R. (1992) *The Rediscovery of the Mind* (MIT Press, Cambridge MA).

Singer, W. (1990) Search for coherence: a basic principle of cortical self-organization, *Concepts Neurosci.* **1**, 1–26.

Stent, G. S. (1981) Strength and weakness of the genetic approach to the development of the nervous system, *Annu. Rev. Neurosci.* **4**, 163–194.

Waddington, C. H. (1969) The theory of evolution today, in *Beyond Reductionism*, edited by A. Koestler and J. R. Smythies (Hutchinson, London).

Zeki, S. M. (1980) The representation of colours in the cerebral cortex, *Nature* **284**, 412–418.

Glossary

Action potential A brief regenerative, all-or-nothing electrical potential that propagates along an axon.

Afferents Axons conducting impulses towards the central nervous system.

Anion A negative ion, such as Cl^-.

Anode A positive pole, to which an anion is attracted.

Antidromic Descriptive of an impulse that travels from axon terminals towards the neuronal soma, i.e. opposite to the usual (orthodromic) direction.

Arachnoid *See* Cerebrospinal fluid.

Arborization The branching of nerve fibres.

Astrocytes *See* Neuroglia.

Attention Concentration on some sensory experience.

Axon The process or processes of a neuron conducting impulses, usually over long distances.

Axon collateral A branch from the main axon.

Axon hillock The region of the cell body at which the axon originates; often the site of impulse initiation.

Basophillic Dark staining of nerve cells.

Bouton A small terminal expansion of the presynaptic nerve fibre at a synapse; the site of transmitter release.

Cable transmission Transmission of a potential charge along a nerve fibre in the manner of an electric current through a fibre and not as an impulse.

Catecholamine A synaptic transmitter substance. *See* Synaptic vesicles.

Cathode A negative pole, to which a cation is attracted.

Cation A positively charged ion, such as Na^+, K^+, or Ca^{2+}.

Caudal Towards the tail or posterior end (in humans the term is 'inferior').

Central nervous system (CNS) The brain and spinal cord.

Cerebellum The smaller (posterior) part of the brain lying behind the cerebrum and responsible for movement control.

Cerebral hemispheres The components of the cerebrum.

Cerebral ventricle A cavity (lumen) in the brain filled with cerebrospinal fluid.

Cerebrospinal fluid (CSF) A clear liquid filling the ventricles of the brain and spaces between meninges, arachnoid, and pia.

Cerebrum The principal (anterior) part of the brain.

Channel An opening or 'pore' in a membrane through which ions or molecules can move.

Cognitive experience A conscious thought or experience.

Conductance The reciprocal of electrical resistance and thus a measure of the ability of a circuit to conduct electricity.

Contralateral Relating to the opposite side of the body.

Convergence The coming together and making of synapses by a group of presynaptic neurons on one postsynaptic neuron.

Coronal section A vertical section through the skull at right angles to the front–back (sagittal) axis, i.e. a plane parallel to the face.

Corpus callosum The tract of nerve fibres that connects the cerebral hemispheres.

Cortical columns The aggregate of cortical neurons sharing common properties, e.g. sensory modality, receptive field position, eye dominance, orientation, movement sensitivity.

Decussation The intercrossing of similar structures.

Dendrite The process of a neuron specialized to act as a receptor; the postsynaptic region of a neuron.

Dendron A composite made by the bunching of the apical dendrites of pyramidal cells.

Depolarization The reduction of a membrane potential from the resting value towards zero.

Divergence The branching of a neuron to form synapses with several other neurons.

Dorsal Pertaining to or situated near the back of an animal.

Efferent Axons conducting impulses outward from the central nervous system.

Electrochemical potential The voltage difference between two solutions insulated from each other; it is derived by the Nernst equation from the logarithmic relationship of the concentrations of the charged particles on either side of the insulating membrane.

Electroencephalogram (EEG) The record taken of the electrical activity of the brain by external electrodes on the scalp.

Electrotonic potentials Localized, graded potentials produced by sub-threshhold currents, determined by passive electrical properties of cells.

Encephalon The brain.

Ependymal cells A layer of cells lining the ventricles of the brain. *See* Neuroglia.

EPSP Excitatory postsynaptic potential in a neuron.

Equilibrium potential The membrane potential at which there is no net passive movement of an ion species into or out of a cell.

Excitation A process tending to produce action potentials.

Exocytosis A process whereby synaptic vesicles fuse with presynaptic terminal membrane and empty transmitter molecules into the synaptic cleft.

Facilitation Greater effectiveness of synaptic transmission by successive presynaptic impulses, usually owing to increased transmitter release.

Fasciculus A bundle of nerve fibres.

Folium Any of the leaf-like subdivisions of the cerebral cortex.

Fusiform Long, tapering at both ends.

Ganglion A collection of neurons that send and receive impulses; many in the autonomic system relay synaptically to the viscera.

Glia *See* Neuroglia.

Golgi type II neurons A short-axoned neuron with an axon that arborizes in dendrite-like fashion in the neighbourhood of the cell body.

Grey matter Part of the central nervous system composed predominantly of the cell bodies of neurons and fine terminals, as opposed to major axon tracts (white matter).

Hippocampus A special primitive part of the cerebral cortex.

Hyperpolarization An increase in membrane potential from the resting value, tending to reduce excitability.

Impulse *See* Action potential.

Inhibition The effect of one neuron upon another tending to prevent it from initiating impulses. Postsynaptic inhibition is mediated through a permeability change in the postsynaptic cell, holding the membrane potential away from threshold. Presynaptic inhibition is mediated by an inhibitory fibre upon an excitatory terminal, reducing the release of transmitter. Electrical inhibition is mediated by currents in presynaptic fibres that hyperpolarize the postsynaptic cell and do not involve the secretion of a chemical transmitter.

Initial segment The region of an axon close to the cell body; often the site of impulse initiation.

Intention Concentration on a proposed movement.

Interneuron A neuron that is neither purely sensory nor motor but connects other neurons.

Ipsilateral On the same side of the body.

IPSP Inhibitory postsynaptic potential.

Mauthner cell A large nerve cell in the mesencephalon of fishes and amphibians, up to 1 mm in length.

Meninges Glial coverings of the brain.

MEPP Miniature end plate potential; a small depolarization at a neuromuscular synapse caused by spontaneous release of a single quantum of transmitter from the presynaptic terminal.

Mesencephalon The lower medial part of the brain.

Modality The class of sensation (e.g. touch, vision, smell).

Motoneuron (motor neuron) A neuron that innervates muscle fibres.

Motor unit A single motoneuron and the muscle fibre it innervates.

Muscle spindle A fusiform end organ in skeletal muscle in which afferent sensory fibres and a few motoneurons terminate.

Myelin The white substance of fused glia ensheathing nerve fibres.

Neocortex The most recently developed part of the cerebral cortex, composing the cerebral hemispheres.

Nerve fibre An axon (the principal branch from a nerve cell) that may extend for long distances.

Neuroglia, or glia Non-neuron satellite cells associated with neurons. In the mammalian central nervous system the main groupings are astrocytes and oligodendrocytes; in peripheral nerves the satellite cells are called Schwann cells.

Neuron (nerve cell) The biological unit of the brain and of the remainder of the nervous system.

Neuron theory The theory that the nervous system is composed of individual neurons that are biologically independent but communicate informationally by synapses.

Node of Ranvier A localized area devoid of myelin occurring at intervals along a myelinated axon.

Noise Fluctuations in membrane potential or current due to random opening and closing of channels.

Nucleus A large basophillic mass, usually centrally placed within the cell body; it contains the DNA that provides the genetic instruction for the cell.

Oligodendrocytes *See* Neuroglia.

Orthodronic *See* Antidronic.

Paracrystalline Crystal-like arrangements of nerve-cell structures.

Pia A glial sheath.

Postsynaptic membrane The nerve cell or other receptor cell membrane immediately related to the synapse formed by presynaptic fibres ending on it.

Presynaptic fibres The terminal branches of nerve fibres that end as synaptic knobs.

Presynaptic vesicular grid (PVG) The presynaptic structure formed of synaptic vesicles and dense projections.

Pyramidal cells The principal neurons of the cerebral cortex of pyramidal shape.

Quantal release Secretion of multimolecular packets (quanta) of transmitter by the presynaptic nerve terminal.

Quantal size The number of molecules of neurotransmitter in a quantum.

Quantum content The number of quanta in a synaptic response.

rCBF Regional cerebral blood flow.

Receptive field The area of the periphery whose stimulation influences the firing of a neuron. For cells in the visual pathway the receptive field refers to an area on the retina whose illumination influences the activity of a neuron.

Receptor 1. A sensory nerve terminal. 2. A molecule in the cell membrane that combines with a specific chemical substance.

Resting potential The steady electrical potential across a membrane in the quiescent state.

Rostral Towards the rostrum or nose; thus, in the anterior direction in the central nervous system.

Sagittal section A section taken parallel to the anterior–posterior plane.

Saltatory conduction Conduction along a myelinated axon whereby the impulse leaps from node to node.

Sensorium The seat of sensation, located in the brain; it is often used to designate the condition of a subject relative to his or her consciousness or mental clarity.

Sensory modality The distinguishing names of all the various sensations arising from diverse inputs wth their specific receptor organs.

Seratonin A transmitter substance. *See* Synaptic vesicles.

Servomechanism A mechanism designed for feedback-control operation.

SMA Supplementary motor area.

Soma The cell body.

Striate cortex The primary visual region of the occipital lobe marked by striation of Gennari, visible wih the naked eye. (Also known as area 17 or visual I.)

Synapse Site at which neurons make functional contact; a term coined by Sherrington.

Synaptic cleft The space between the membranes of the presynaptic and postsynaptic cells at a chemical synapse across which transmitter must diffuse.

Synaptic plasticity The property of synapses whereby they are changed in functional efficiency, probably by virtue of changes in size.

Synaptic vesicles Small membrane-bounded sacs contained in presynaptic nerve terminals. Those with dense cores contain catecholamines and serotonin; clear vesicles are presumed to be the storage sites for other transmitters.

Threshold 1. The critical value of membrane potential or depolarization at which an impulse is initiated. 2. The minimal stimulus required for a sensation.

Transmitter A chemical substance liberated by a presynaptic nerve terminal causing an effect on the membrane of the postsynaptic cell, usually an increase in permeability to one or more ions.

Trophic influence An action from one part of a cell to another, or from one cell to another, that is concerned with the growth, maintenance, and metabolism of the cell.

Ventral Towards the under or belly side of the body.

Ventricles Cavities within the brain containing cerebrospinal fluid and lined by ependymal cells.

White matter Part of the central nervous system appearing white; consisting of myelinated fibre tracts.

Index

Springer-Verlag and the Environment

We at Springer-Verlag firmly believe that an international science publisher has a special obligation to the environment, and our corporate policies consistently reflect this conviction.

We also expect our business partners – paper mills, printers, packaging manufacturers, etc. – to commit themselves to using environmentally friendly materials and production processes.

The paper in this book is made from low- or no-chlorine pulp and is acid free, in conformance with international standards for paper permanency.